Lecture Notes in Economics and Mathematical Systems

Managing Editors: M. Beckmann and W. Krelle

233

Gerhard Wagenhals

The World Copper Market

Structure and Econometric Model

Springer-Verlag
Berlin Heidelberg New York Tokyo 1984

Author

Dr. Gerhard Wagenhals
Alfred Weber-Institut, University of Heidelberg
Grabengasse 14, D-6900 Heidelberg, FRG

cc

ISBN 3-540-13860-9 Springer-Verlag Berlin Heidelberg New York Tokyo
ISBN 0-387-13860-9 Springer-Verlag New York Heidelberg Berlin Tokyo

Library of Congress Cataloging in Publication Data. Wagenhals, Gerhard. The world copper
market. (Lecture notes in economics and mathematical systems; 233) Bibliography: p. 1. Copper
industry and trade–Mathematical models. I. Title. II. Series.
HD9539.C6W34 1984 338.2'743'0724 84-20198
ISBN 0-387-13860-9 (U.S.)

Printing and binding: Beltz Offsetdruck, Hemsbach/Bergstr.
2142/3140-543210

ACKNOWLEDGEMENTS

The advice of Professor F. Gerard Adams in the preparation of this report is very gratefully acknowledged. His helpful suggestions and criticisms were invaluable. Without his guidance I could not have finished this study.

Thanks are also due to all who have assisted with this report in any way, through practical help, correspondence, useful suggestions and a readiness to supply information. Especially I would like to thank Professor Jere R. Behrman, Professor J.D.A. Cuddy, Dr. Wolfgang Gluschke, Professor Lawrence R. Klein, Professor Peter Pauly, Dr. Walter Sies and Dr. Kenji Takeuchi.

Professor Malte Faber, Professor Hans Jürgen Jaksch, Professor Jürgen Siebke and an unknown referee thoroughly read earlier drafts of this study. They made many very helpful suggestions to improve its contents and presentation.

The Deutsche Forschungsgemeinschaft very generously provided financial support for a two-year visit at the Economics Research Unit of the University of Pennsylvania, Philadelphia, where the main part of the research for this book was done.

I would also like to express my deep appreciation to Chris Swan, who not only corrected my English, but who always encouraged me during my visit in Philadelphia.

Certainly, this study would not have been possible without many helpful discussions and the steady support of my wife, Doris.

CONTENTS

LIST OF FIGURES

LIST OF TABLES

ABBREVIATIONS

BLS	U.S. Bureau of Labor Statistics
C	Cumulated percentage share of full CIPEC members
C(1)	Percentage share of the largest unit
C(3)	Cumulated percentage shares of the three largest units
C(6)	Cumulated percentage shares of the six largest units
CIPEC	Intergovermental Council of copper exporting countries
CODELCO	Corporacion Nacional del Cobre de Chile
COMEX	Commodity Exchange, Inc., New York
CRA	Charles River Associates, Boston, Mass.
CRU	Commodities Research Unit, London and New York
GDP	Gross domestic product
GNP	Gross national product
lb.	Pound sterling
LIBOR	London interbank offering rate
LME	London Metal Exchange
n.a.	Not available
OLS	Ordinary least squares
SUR	Seemingly unrelated regressions
ton	Metric ton
tpy	Tons per year
ppm	Parts per million
WBMS	World Bureau of Metal Statistics
WEFA	Wharton Econometric Forecasting Associates

Chapter 1: Introduction

1.1 The Importance of Copper

Copper, the red metal, has been known in history for thousands of years. It may have been mankind's first metal (Joralemon: 1973). And still, probably more than one hundred decades after native copper was used for the first time (Muhly (1973: 171)), today, copper is a very important commodity:

1. Only aluminum (first in 1963) surpasses refined copper in terms of the total world's mine production and consumption. It outpaces zinc, lead, nickel and tin[1].

2. Refined copper is one of the most important export products of the developing countries.
   In 1975, refined copper ranked 8th in the developing countries' export values in general, it was 6th among their non-fuel exports, and their most important export commodity among the non-ferrous metals[2].

3. Many small and medium sized industrialized countries depend heavily on copper imports.
   For example, West Germany's share in world mine production has always been smaller than 0.1 per cent. In the last few decades, however, the Federal Republic's consumption share has amounted to some 8 % in 1982.

4. Copper is of utmost importance for the export earnings of several countries.
   The ratio of copper export earnings to total export earnings in 1982 was
   - 89 % for Zambia,
   - 53 % for Papua New Guinea,
   - 45 % for Chile,
   - 40 % for Zaire,
   - 14 % for Peru,
   - 14 % for Botswana, and
   - 6 % for the Philippines[3].

5. The price of copper is highly volatile.

For the period from 1950 to 1979 the coefficients of price variation (after accounting for inflation) at the London Metal Exchange were 33 % for copper, 35 % for zinc, and 27 %, 18 %, 14 % for lead, tin, and aluminum respectively.

The impact of all these fluctuations on the export earnings of the developing countries is considerable, and the consequences can be quite drastic. The impediment to continuous development planning is just one example. On the other hand recent history has shown that macroeconomic repercussions of commodity price rises can also be serious for industrialized countries.

These few remarks give evidence of the importance of copper in the world economy and they justify a more thorough study of this market[4].

1.2 Objectives and Overview

The general purpose of this book is to identify the saliant market forces and the past behavior of the world copper industry, based on economic theory.

Specifically, its main objectives are to describe the most important patterns of copper production, consumption, price formation and trade, and to assess the future availability of the red metal.

The study uses two approaches. Part I describes the world copper market verbally, based on detailed statistical facts. Part II relies on mathematical economics and on econometric theory in an attempt to capture some of the most important behavioral aspects of different market participants discussed in Part I.

The study is organized as follows.

Chapter 2 describes the production side of the world copper market. The processing of the metal, the development of the copper mine, smelter and refined production, and their concentration in the last few decades is investigated in some detail. The chapter closes with an examination of the copper scrap market.

Chapter 3 deals with the consumption side of the market. After a description of the most significant developments after the second World War, the technical and economic determinants of substitution are assessed.

Chapter 4 surveys the patterns of copper trade. It describes the development and achievements of the organization of the most important copper exporting developing countries (CIPEC). Then, the pricing of copper and the working of the London Metal Exchange are explained. The existence of a loosely oligopolistic North American copper market and its relation to the world market is stressed.

Chapter 5 looks into copper as an exhaustible resource. After an appraisal of copper reserves and resources yesterday and today, the perspectives of future copper availability are assessed. The scarcity of the red metal is measured in terms of various indicators, updating several previous studies of this topic.

Based on the first part of the book, which describes and analyzes the world copper industry statistically and verbally, the second part presents a new econometric model of the world copper market. To begin with, chapter 6 surveys earlier quantitative analyses of the copper industry, including not only econometric approaches, but also engineering and other studies. Based on this research, and incorporating several new aspects, an econometric world copper market model is developed in the following chapters.

Chapters 7, 8, and 9 deal with the specification and estimation of this model. Derived from the body of economic theory and considering the special features of the world copper market described in earlier chapters of this book, the market's most outstanding quantitative features are summarized in some 60 behavioral equations. Figure 1-1 shows the main relationships of the model.

Chapter 7 specifies equations for the primary supply and for the mine production capacities in eight countries and two supranational regions. Unlike traditional models, which use a conventional partial adjustment approach, this study assumes that copper producers maximize their profits given their capacity restrictions. Equations for mining capacities, which have been neglected in most econometric copper market models, then are derived from the hypothesis that producers optimize their discounted net cash flow. Typical equations are described formally. The most important estimation and simulation results are explained and interpreted.

Chapter 8 deals with the specification and estimation of behavioral equations describing the total use of the metal, the demand for refined copper and the demand for storage. To account for a common covariance structure of the estimated residuals due to similar business cycles in most copper consuming countries, Zellner's method of seemingly unrelated regressions has been used to estimate the parameters of the consumption equations. The derivation of the speculative inventory behavior takes into regard the efficiency of the London Metal Exchange's copper price formation in the short run and applies a rational expectation hypothesis. This chapter also describes and interprets the most important estimated equations.

Chapter 9 comprises the rest of the model. It includes the specification and estimation of equations for the secondary supply in the most important copper consuming countries and for several prices. The determination of the world market spot price follows a dynamic stock disequilibrium approach. For the first time, this model also explicitly includes the futures market at the London Metal Exchange.

In chapter 10, the validity of the model is tested by dynamic simulation techniques. It is then used to perform a set of eight policy simulations. The chapter concludes with an outlook on uses of the model in price forecasting and as a tool in policy making and planning. It references to studies that have been published elsewhere.

The book finishes with several appendices. They include a description of the data used, and a complete list of the estimated equations. An extensive bibliography completes this book.

1.3 Notes

1. Unless otherwise noted, all quantitative statements in this report are derived from data compiled in various issues of Metal Statistics published annually by the Metallgesellschaft A.G., Frankfurt. Experts for non-ferrous markets generally acknowledge that this publication is the best source for copper data. The preliminary figures for 1983 are quoted or calculated from Mineral Commodity Summaries (1984).

2. Calculated from UNCTAD's Handbook of International Trade and Development Statistics (1979).

3. Calculated from International Financial Statistics published by the International Monetary Fund, Vol. 37(3), March 1984.

4. In view of copper's salient role, it is not surprising, that a broad spectrum of literature evolves, concerning the economic, technological and natural aspects of the metal. Wagenhals (1983a, Chapter 1) briefly surveys recent literature on the world copper industry and trade in general. More detailed references follow in all later chapters of this book.

Figure 1-1:

Simplified structure of the world copper model

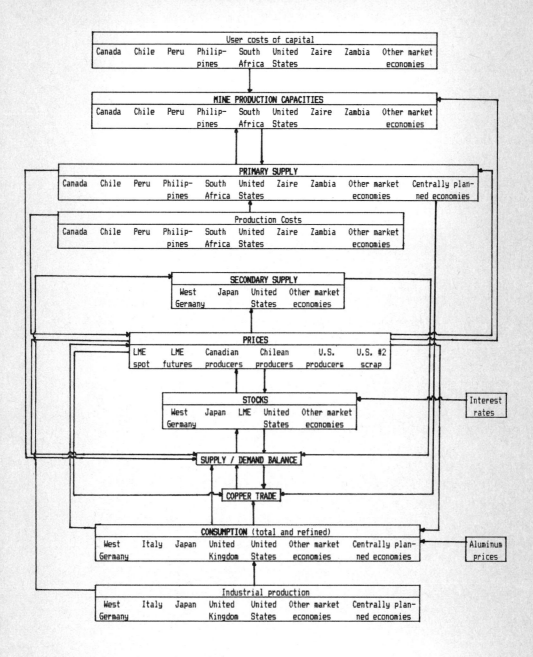

PART I

Chapter 2: Production

   This chapter deals with the production patterns of the red metal. First, a few
non-technical remarks about the art of copper processing present the metallurgical
background necessary to understand the detailed statistical description of the
development of the various stages of copper production following in the later
sections. Like all chapters in the first part of this book, chapter 2 provides the
illustrative basis for the development of the econometric model in the second part
of the book. But, moreover, it also intends to stress factors which have not been
accounted for by the econometric model, and thus to remind the reader of its
limitations.

2.1 Metallurgical Background

   To understand the world copper industry, a certain knowledge of copper
processing is indispensable. The most important facts are listed in this section.
   The technical process of copper-making from ore extraction to the production of
the marketable, almost pure metal consists of several steps:

      - mining,
      - concentration,
      - smelting, and
      - refining.

   Mining, the extraction of the copper bearing ore from the earth, is the first
processing step. 90% of all copper is mined from sulphide minerals, the rest from
oxidized minerals. Both types of minerals require different treatment[1].
   Sulphide ores are crushed, ground and then concentrated by physical techniques,
mainly floatation, to obtain concentrates with 10% to 40% metal content. During the
smelting phase the concentrate is melted in a furnace to produce so-called matte,
which contains between 30% and 55% copper. The liquid matte is poured into a
converter and oxidized to remove sulphur and other impurities, leaking some sulphur
dioxide into the atmosphere[2]. This converting produces so-called blister copper,
containing between 99.5% and 99.8% copper. In the refining stage, by an
electrochemical process, copper is dissolved from anodes and plated onto cathodes
with almost no impurities. This leads to electrolytic cathode copper with at least

99.99% metal content. Fire refining plays only a very small role, because it is less effective in the removal of impurities.

Oxide ores are generally treated by hydro-metallurgical methods. Broken, crushed and ground oxide ores are leached with sulphuric acid. For low grade ores, copper is recovered by cementation on scrap iron. If the leach solution is strong, then copper can be recovered by electrowinning, yielding cathodes somewhat less pure than electrolytic cathodes.

Irrespective of the metallurgical techniques used, refined copper always contains at least 99.9% pure metal. At primary refineries, it is cast in cathodes (61%), wirebars (23%), billets (6%), cakes (5%), ingots and ingot bars (3%) and other forms (2%)[3].

Various improvements of the metallurgical processes have achieved noteworthy economic significance in recent years. For example, processes of single-step copper making have reached commercial applicability. Blister copper is produced directly from sulphide concentrates, omitting the intermediate steps. The main advantages of such continuous processes are more efficient energy useage and the production of a continuous gas stream. This allows an easier sulphur dioxide removal than the traditional method[4].

Another metallurgical process which has gained rapidly in importance in the last decade is the continuous casting of copper bars. In wire and rod manufacturing especially continuous casting / rod-rolling systems replace more and more the casting and remelting of individual wirebars[5].

This very short summary of the metallurgical processes in the copper industry should be a sufficient background for the detailed description of production patterns, which is the subject of the following sections.

2.2 Mine Production

Mine output, copper extracted from ore, is one of the most important variables explained by the econometric model in Part II. This section verbally and statistically describes the development of copper mine production and its distribution on various countries in the last few decades.

First, however here is a factual summary.

Basic Facts

The United States, the Soviet Union and Chile are the world's largest copper producers. While the United States' share has declined gradually since the second World War, the centrally planned economies' proportion has grown, mainly due to a large increase in Poland's copper output. The developing countries' share peaked in the early 1960s. Since then it has decreased due to the reluctant investments of multinational companies. Many of the important copper exporting developing economies have undergone a series of nationalizations.

The concentration of mine production within the market countries has declined. Low grade copper deposits can be found in many places all over the world. Technical progress has rendered possible the mining of these marginal deposits. Additionally, the change in the ownership structure of the Western World's copper mines has contributed to this trend. Immediately after the second World War, only a few multinational companies owned most of the market countries' copper capacity. Today the governments of many developing countries own or have majority shares in the copper enterprises within their borders. The next few sections give more quantitative information and substantiate these assertions.

Mine Production

The world mine production in 1982 reached 8.2 million tons of copper (metal content of ores) in contrast to 7.1 million tons ten years earlier. The Western World's share amounted to 76% (in 1972: 80%).

Table 2-1 lists the most important copper ore extracting countries.

Table 2-1:

Shares in the total world's mine production of copper ore (metal content), in per cent

Country	1950	1960	1970	1980	1982
United States	32.7	23.1	24.4	15.0	13.9
Soviet Union	8.6	11.8	14.4	14.4	14.4
Chile	14.4	12.6	10.8	13.6	15.1
Canada	9.5	9.4	9.5	9.1	7.4
Zambia	11.8	13.6	10.7	7.6	6.4
Zaire	7.0	7.1	6.0	5.8	6.1
Peru	1.2	4.3	3.4	4.7	4.3

Between two and five per cent of the world mine output in 1982 was extracted in Poland (5%), the Philippines (4%), Australia, Mexico, the Republic of South Africa (3% respectively), Papua New Guinea and the People's Republic of China (2% in each). Yugoslavia, Indonesia, and Japan play a small role with some 1% of the world mine production respectively. The European Community's output share did not reach one per cent.

From 1969 to 1979 Poland's copper output grew sevenfold. The Philippines' production nearly doubled in the same period. Indonesia and Papua New Guinea began their mine production in 1972.

In 1982 and 1983, Chile's copper output surpassed U.S. mine production. This was, however, an extraordinary event. For these two years, the U.S. mining sector only worked at some two-thirds capacity, while Chile's copper mines operated almost at full capacity.

Concentration

The market power of privately owned firms fell dramatically since the second World War. In 1948 the seven largest privately owned copper enterprises shared 70% of the market economies' production. Thirty years later, their production accounted for less than 25% [6].

The country specific concentration of copper production also decreased, although less spectacularly. Figure 2-1 shows some concentration indexes C(i) [7], i.e. the cumulated percentage shares of the i largest copper producing countries, since the early 1950s. C denotes the cumulated percentage share of the countries joined in the "Conseil Intergouvernemental des Pays Exportateurs de Cuivre" (CIPEC).

The main reasons for the observed decline in concentration at the mining stage are growing copper reserves and resources in many parts of the world, wide diffusion of technology for the working of low-grade porphyry ore bodies, backward-integratation of some Japanese and German companies, and expanding operations of state-owned copper mining enterprises[8].

Dropping concentration ratios indicate an increasing competitiveness in the world copper market. This fact supports the use of a competitive approach in the derivation of the primary supply and mine production capacity equations in Part II of the study.

Figure 2-1:

Concentration ratios in copper mining, 1950-1983

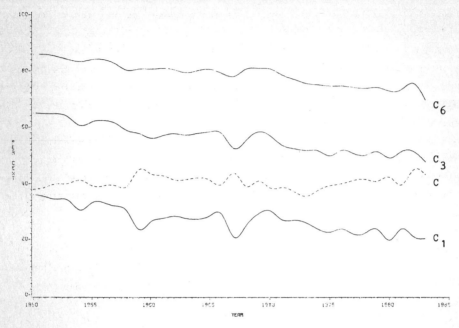

Exogenous Shocks

Basic economic theory suggests that relative prices influence copper supply. Naturally, this will show up in the estimating copper mine production equations. The next section stresses the importance of mine capacity as a limiting factor of actual ore throughput. But apart from these endogenous variables, which of course are explained in the framework of our econometric model, other factors sometimes influence the current level of metal output.

Look again at Figure 2-1. In the years 1954, 1959, 1967/68, and 1980, it shows four slumps in the share of the largest copper producer, i.e. - until recently - the United States' share. These troughs have been caused by serious labor strikes at leading U.S. copper producing companies. Such exogenous shocks have sometimes lead to considerable variations in the short run primary copper supply. Take for example the 1967/68 downturn in U.S. copper output: Although the copper industry had expected the 1967 strike and built up sizable stocks, it turned out to be the largest and costliest walkout in the history of copper mining. It closed 90% of the U.S. copper mines. Operations were not resumed until after April 1968 [9]. This and similar extraordinary events are not explained by our econometric model. Dummy variables account for such incidences.

Strikes, outlocks, and other political factors influence mine production not only in the United States, but also elsewhere. The secession of the former Katanga province, for example, affected Central Africa's copper industry for a long time. Not only because of the hostilities and the war itself, but also because of the loss of skilled labor necessary to operate and maintain equipment after the shut-down of Union Minière in December 1962. The closure of the Benguela railroad in 1975 as a consequence of the civil war in Angola aggravated Zambia's already difficult logistical problems. These examples of political disruptions influencing copper output could easily be expanded.

Not only political events, but also economic developments in other industries have influenced the copper market. This in especially pronounced in Canada, where several times copper output has been reduced due to cutbacks in general mining, where copper is only a by-product. Production cut-backs due to flagging nickel demand (1958, 1977) or because of a strike at the world's largest nickel producer (1978) have reduced Canada's copper output substantially.

Apart from human actions, natural disasters (such as floods and land-slides, fires and cave-ins) have impeded the production of the red metal. Claiming the lives of many miners, the collapse of Zambia's Mufulira underground mine in September 1970 is the most tragic example.

Thus, exogenous shocks like wars, labor strikes, and similar factors hinder or even cut off copper production. Although the stochastical error term in the behavioral equations of our model accounts for small, non-systematical effects, major breakdowns are accounted for by exogenous dummy variables.

2.3 Mine Production Capacities

Actual throughput of copper ores is restricted by the capacity of a mine[10]. Therefore the mine production capacity of a country significantly determines its actual output. Thus the following econometric model explains not only copper production levels, but also the capacities in many countries and regions. After summarizing basic facts about copper mine production capacities, this section delineates some factors influencing the expansion of old and development of new production capacities in the world copper industry, including some information about the metal's most important by-products[11].

Basic Facts

From 1970 to 1982 the world's copper mine capacities grew from some 6.5 million tons to more than 9.7 million tons. Its distribution among countries is similar to the allocation of the mine production.

Capacity utilization, in terms of mine output per mine capacity, averaged 89% for the Western World from 1960 to 1980. Among the major copper producers it was highest in South Africa (98%) and the Philippines (94%), and lowest in the United States (83%) and Canada (87%). Since 1970, U.S. mine production has not. attained attained a capacity utilization rate above 90%. CIPEC's share in the market economies' total capacity fell almost ten percentage points in this period. Like mine production concentration, capacity concentration also declined due to the factors listed in the previous section.

Long- and Short-term Expansions

Additional mine capacity is installed by starting up new mines or by expanding existing pits.

The time from preliminary exploration to the beginning of extraction at a new mine can be rather long. In a case study for the U.S. Bureau of Mines, Burgin (1976) found that in selected Arizona copper mines the exploration time varied between 1 and 15 years, and the construction time ranged from 1 to 8 years. Up to two years are required for the construction of a benefication plant. In developing countries, the gestation period can be considerably longer due to the necessity of time consuming infrastructure investments and due to potential delays because of bargaining problems between host countries and foreign investors. But even after a mine's official start, it can take a few years, before it operates at full capacity[12]. For example, one of the world's largest copper occurences, the Gunung Bijih deposit in Indonesia was discovered by the Dutch East Borneo Company in 1936. The initial production began in late 1972. The mine reached its full capacity only in 1974 [13].

It is clear that decisions concerning such large scale mining projects inter alia depend on long-term expectations with regard to the future performance of the world copper market, especially on desired production, on anticipated costs and prices, as well as on the long-run, non-economic goals of government-owned mining enterprises in some developing copper exporting economies.

However, this long run aspect should not veil the significance of short-term expansions, i.e. the extension of already existing pits. Gluschke et al. (1979: 67) estimate that in "the past, the ratio of new installations to expansions has been about 60:40". According to these authors, the Kennecott Company alone could add up to 150,000 tpy over a few years[14]. The gestation period of such expansions lies between 18 and 36 months; investment costs are smaller, and economies of scale can lead to further cost advantages compared to new mines going on-stream[15]. Also, expansion programs can easily be suspended pending improvement of the copper market, and be put on a stand-by basis. Thus, a significant part of the actual capacity changes depends on a short-run assessment of the copper companies, and therefore strongly on their short-run appraisal of prices and desired quantities.

Whether an existing mine is scheduled to be expanded or whether a new mine is projected to go on-stream, the realization of these plans depends on many factors apart from prices and desired production. One of the most important is the availability of capital, a problem for most of the copper producing developing countries, where capital markets are usually only rudimentary[16]. Such financial constraints have played an important role in limiting capacity expansions by state mining enterprises in less developed countries. For example, Minero Peru has been seeking capital for a number of years to develop several ore bodies without much success. Due to lacking consistent time series data our econometric model allows for financial constraints only implictly in the form of unit costs of mining and user costs of capital indices. Therefore, an explanation of the financing of LDC capacities follows.

Before the 1960s, equity contributions from transnational corporations and internal cash generation typically financed new mining projects in the developing countries. Economies of scale in the mining of low-grade porphyry ores required increasing capital inputs to reach optimal mine sizes. Investment financing shifted gradually to loans from commercial banks, often syndicated on a broad world-wide base, and to credits from copper consumers, tied to long-term supply arrangements. Today, contributions from international organizations help to catalyze the funds and to counteract the decline in direct foreign investments.

The Philippines is the only developing country, where domestic private capital makes an important contribution to the financing of copper mining projects. Among the state mining enterprises only the Corporacion Nacional del Cobre de Chile (CODELCO) generates sufficient cash flow to contribute significantly to investments. All other copper producing developing countries have to rely on international capital markets, especially on the Eurodollar market. Interest rates offered to finance mining projects are 0.5% to 1.0% above LIBOR, the London interbank offering rate. This variable is also included in our econometric model. It is one of the determinants of the user costs of capital for many copper producing countries.

Prices, quantities, costs and availability of capital are important determinants of the capacity level, but naturally not the only ones. Other variables, which
our econometric model explicitly takes into account, include depreciation charges,
corporate tax rates and depreciation methods used. Nevertheless, there are some
aspects of the determination of investment in new capacity which do not appear to be
captured completely in the capacity equations. Among others, by-product charges and
credits have to be considered.

By-Products

Many copper deposits bear other substances, whose expected recovery can
influence investment decisions. A notable example is the Ok Tedi mine in Papua New
Guinea, which shall only go on-stream because of high gold by-product credits[17]. The
red metal's most important by-products (in terms of credits as percentages of
production costs) are gold, silver, nickel, zinc, cobalt and molybdenum. Also
platinum group metals, arsenic, rhenium, selenium, palladium, tellurium and uranium
are sometimes recovered together with copper[18].

The occurence of by- and co-products depends on the geographical distribution
of the deposits. Gold plays a major role as copper by-product in Papua New Guinea
and in several South African and Philippinean mines. Santo Thomas II, for example,
one of the Philippines' largest copper mines, is this country's largest gold
producer[19]. In Central Africa, cobalt is the most important by-product. Zaire's and
Zambia's production totalled up to 52% and 18% of the Western World's cobalt output
in 1983[20]. Cobalt and copper production are closely linked by the processing of
cupro-cobalt ores. Molybdenum and metal doré (containing 98.5% silver and 1% gold)
are Chile's most valuable by-products[21]. In Canada's Pre-Cambrian shield, copper
often is jointly produced with nickel or is just a by-product of nickel mining.
Several examples in the last section showed how this influenced the Canadian copper
industry in the past. Apart from nickel, zinc and molybdenum are Canada's most
important copper by-products[22].

Although the existence of by- and co-products often influences the decision
whether a specific deposit is scheduled for start-up, regresson experiments revealed
that generally prices of by- or co-products only insignificantly determine copper
output and capacity decisions on the country level. Therefore the econometric model
in Part II does not allow for these prices. Their weight at a site specific level
should be considered.

2.4 Smelter and Refined Production

The econometric model in Part II includes an estimating refined copper output equation. This section presents some general information about the patterns of refined copper production and about its preliminary stage, smelter production.

Smelter Production

Copper ore is usually beneficated at or nearby the mine due to high transportation costs. Economies of scale require smelters with a relatively high concentrate input. Thus, mine and smelter production are often geographically separated, not only within a country, but also, in a division of labor, between countries[23].

Japan, for instance, is often called the "developing countries' custom smelter". Although her copper mine production remained almost constant in the first 25 years after the second World War and then even declined, her smelter output increased not only due to imports from developing countries, but also from Canada and the United States. The centrally planned economies' smelter output also grew continuously, in contrast to the United States' share, which declined relative to the total, and relative to the market economies' production. Since the mid-1960s the level of the U.S. smelter output almost stagnated. New environmental requirements since 1970 increased U.S. smelting costs, so that - as an extreme example - in 1980 one of the largest primary copper companies closed one of its smelters and shipped the ores to Japan instead of complying with the regulations[24].

The developing countries' and Canada's share in the world's smelter output remained constant, by and large. An increase in some of the developing countries' smelting capacities (notably Zambia's), was compensated by new entrants in the world copper market without smelting capacities (e.g. Indonesia and Papua New Guinea), who shipped their concentrates to Japan and West Germany. In Europe, only the Federal Republic, Poland, Yugoslavia and Spain possess smelting capacities worth mentioning.

After this short summary, a more detailed look at the statistical facts.

The smelter production of copper totalled up to 7.8 million tons in 1982 (1972: 7 million tons). The Western World's share amounted to 76% compared with 80% a decade earlier.

Table 2-2 shows the most important producers of smelter copper.

Table 2-2:

Shares in the total world's smelter production of copper,
in per cent

Country	1950	1960	1970	1980	1982
United States	36.2	26.0	23.6	13.2	12.5
Soviet Union	8.6	11.6	14.6	15.0	15.6
Chile	13.7	11.7	10.2	12.5	13.4
Japan	1.5	4.4	7.9	11.6	12.1
Zambia	11.1	13.4	10.8	7.9	7.4
Zaire	6.8	7.0	6.1	5.6	6.0
Canada	8.3	8.4	7.1	6.4	4.7

Between four and two per cent of the world's smelter output in 1982 was produced in Peru and Poland (4% respectively), the People's Republic of China (3%), the Republic of South Africa, Australia and the Federal Republic of Germany (2% respectively). With some 1% of the world's mine output each, Yugoslavia, Spain and Mexico were only minor producers of blister copper. The European Community's share in the world smelter production added up to no more than 2% and it was concentrated in West Germany. Output grew especially fast in Peru, where smelter output almost doubled between 1970 and 1980, but above all in Poland, where it increased more than five times in this period.

Currently, more countries begin or schedule to start a domestic smelter production[25]. A recent example is Iran, which in 1983 began to smelt domestic copper ores at Sar Chesmeh.

In developing countries, environmental regulations tend to be rather lax and encourage smelting there. Whether in the long run the world's smelter capacity will increase considerably is questionable, however, because of the growing implementation of single-step copper making processes.

Concentration of Smelter Production

Figure 2-1 made it clear and a statistical analysis shows likewise a declining time trend in the concentration of the world's copper mine production. The same trend can be ascertained for the smelter production of copper. This is indicated by table 2-3, where, as above, C(i) denotes the cumulated percentage share of the i largest producers (for i=1, 3, 6), and C stands for CIPEC's cumulated percentage share.

Table 2-3:

Concentration indexes for the market economies' smelter copper output,
in per cent

Year	C(1)	C(3)	C(6)	C
1950	40	67	81	36
1960	31	60	83	42
1970	29	55	81	37
1980	17	49	75	40
1983	16	48	70	37

The world's smelter output is more concentrated in the industrialized countries than the mine production. The following section shows that the developing countries play an even smaller role in the production of refined copper.

Refined Production

In 1982, 9.5 million tons of refined copper were produced world-wide, 75% of that in the market economies (1972: 8.1 million tons, and almost 80% of that in the Western World).

The following table shows the most important producers:

Table 2-4:

Shares in the total world's refined production of copper,
in per cent

Country	1950	1960	1970	1980	1982
United States	42.3	32.9	26.8	17.9	17.6
Soviet Union	8.8	12.2	14.2	15.4	16.0
Japan	2.7	5.0	9.3	10.8	11.3
Chile	9.4	4.5	6.1	8.6	9.0
Zambia	2.5	8.1	7.6	6.4	6.2
Canada	6.8	7.6	5.3	5.4	3.3
West Germany	6.2	6.2	5.3	4.0	4.1

Belgium/Luxemburg produced 5%, Poland 4%, the People's Republic of China 3%, Peru, Australia, Spain, Zaire 2% (each), Yugoslavia, South Korea and Mexico somewhat more than 1% (respectively) of the world's refined copper output in 1982. The European Community's share totalled up to 11%.

A comparison of table 2-4 with tables 2-1 and 2-2 reveals immediately that the industrialized countries' share in the refined production is higher than its share in the world's smelter output.

Concentration of refined copper production in the market economies declined. Table 2-5 presents some evidence. Against this general trend, CIPEC's share C increased in the last 30 years. This is almost only due to Zambia's contribution. She refined just 27% of her copper output in 1950, but all of it today.

Table 2-5:

Concentration indexes for the market economies' refined copper output, in per cent

Year	C(1)	C(3)	C(6)	C
1950	47	65	84	17
1960	39	58	76	19
1970	33	54	76	21
1980	24	50	71	25
1983	22	48	68	25

Finally, figure 2-2 displays the development of the total world's copper mine production (QMTW) and refined production (QRTW) since the beginning of the twentieth century. Because smelter output is roughly as high as the mine output, it is not depicted here. The strong crises between the wars, the end of World War II and the recessions in 1975 and the early 1980s clearly show their marks.

Figure 2-2:

Copper world mine and refined production since 1900

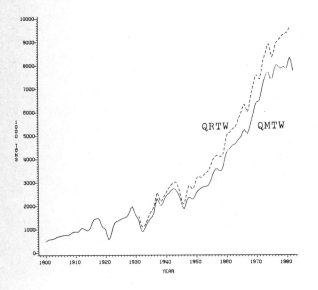

2.5 Scrap

Basic Facts

The differences between the figures of smelter and refined production are due
to the recovery of copper from scrap.

Apart from copper in alloys with ordnance and dissipative uses and from copper
in chemicals, the recycling of the metal is possible in principle[26]. Copper
recovered from scrap is called **secondary copper**, in contrast to **primary copper**,
which is produced from ore[27]. Secondary copper can be won from used objects (**old
scrap**), and is salvaged from outworn, obsolete equipment and machinery. Examples are
discarded wires, rods, pipes or transformers[28]. But secondary copper can also be won
from refuse while manufacturing copper bearing products. Examples for this **new scrap**
are defective castings, punchings or skimmings.

Several types of scrap can be distinguished:
- Scrap which can be remelted and cast directly, with the quality level of cathode
  copper ("#1 wire, heavy scrap").
- High grade copper scrap contaminated by other metals, which needs remelting and
  removal of impurities ("#2 wire, mixed heavy and light scrap").
- Copper alloy scrap (mainly brass and bronze scrap), which can be remelted and
  recast into new alloys, and
- other scrap consisting mainly of low-grade scrap, which has to be remelted and
  re-refined before it can be used again.

According to their copper content and their degree of alloying these types of
scrap can be used as inputs in refineries and foundries, and also sometimes directly
in the production of semifabricated products. In 1980, of 100 kg scrap used in the
United States, 19 kg consisted of #1 and 22 kg of #2 scrap, 34 kg were composed of
various types of copper alloy scrap, including 17 kg yellow brass, the rest was
low-grade scrap and residues[29].

Copper produced from scrap is virtually a perfect substitute for primary copper. Some differences in the economics of secondary copper, however, do exist. These
are:
- **Costs:** The costs of producing secondary copper depend on the grade and quality of
  the scrap input. Because of a simpler technology, capital and processing costs for
  high-grade, good-quality scrap are smaller compared to primary copper[30].
- **Energy requirements:** The energy requirements to recycle copper scrap are smaller
  than the energy needs for primary copper production. Various sources indicate that
  scrap recovery demands three to twenty times less energy per ton of refined
  metal[31].

- **Market structure:** Copper scrap markets are competitive in all countries, but the primary copper market in North America shows some oligopolistic features[32].
- **Prices:** Reflecting these differences in market structures, different pricing systems for primary and secondary copper have evolved in North America. In Europe, scrap prices are in line with the London Metal Exchanges' prices[33]. This free market price leads the European scrap price in the short run[34].

Some more evidence about scrap production and its role in the world copper market follows.

Secondary Production

In the market economies, 1.2 million tons of copper were refined from scrap in 1982. Most of the scrap was processed in the United States (39%), the European Community (31%; more than half of it in the Federal Republic) and in Japan (11%).

In the developing countries only a few facilities for secondary production exist. Yugoslavia, the Republic of Korea, Brazil and Mexico produce some copper from scrap, but in 1980, they together did not even reach 10% of the market economies' total secondary production.

Table 2-6 gives some more detailed figures about the most important secondary copper producers in the last three decades.

Table 2-6:

Shares in the market economies' secondary refined copper production, in per cent

Country	1950	1960	1970	1980	1982
United States	44.1	39.0	37.6	36.9	38.6
West Germany	23.8	19.4	17.5	14.5	15.3
Japan	9.6	9.0	8.9	9.7	11.0
United Kingdom	14.5	15.7	13.6	7.2	6.1
Belgium	n.a.	n.a.	6.2	5.4	5.2

In the last decade, the total amount of scrap used in the Western World increased only slightly from 3.3 to 3.4 million tons. Manufacturers in the market economies used 2.3 million tons of scrap directly in 1982, 30% of that by the United States, 30% by the European Community and 20% by Japan. The share of the above-mentioned developing countries reached only 4%.

Table 2-7 shows the share of scrap input in the total refined copper production for selected countries and years.

Table 2-7:

Shares of secondary production in the total refined copper production of selected countries,
in per cent

Country	1950	1960	1970	1980	1983 [35]
West Germany	59.8	42.6	48.9	49.9	40.4
United Kingdom	37.3	48.5	76.0	40.7	52.9
Japan	56.6	24.5	14.5	12.3	11.8
United States	16.3	16.1	21.3	28.0	27.1
Market economies	17.4	16.2	18.7	17.1	15.7

Obviously, a time trend common to all countries cannot be found[36]. Only in the United States scrap utilization grew significantly.

Among the market economies, the scrap recovery rate is highest in the Federal Republic and in the United Kingdom. Because the average Western World's copper recycling rate has been rather small compared to these countries, it might be asked, whether there is scope for an increased copper recovery from scrap in the near future.

In an optimistic study, Radetzki and Svensson (1979) suggest that, if the demand for copper contracts by one per cent annually, old scrap could satisfy this demand completely. Their analysis is based on some very simplifying assumptions, but it is made clear that effective recycling can increase the availability of the metal.

However, it is questionable, whether the scope for additional recycling in the copper industry is really large. In the last thirty years the efficiency of copper recovery remained constant. Chapman (1974: 317) estimates an 85% efficiency in the recovery of new scrap and a 45% efficiency in the recovery of old scrap in the United States. These percentages surpass the corresponding figures for aluminum. Gluschke et al. (1979: 98-100) also note that, compared to other metals, copper's recovery rate is already high. The major part of unrecycled scrap arises in household waste, but it contains only a relatively small amount of copper with a very low grade. This confirms Gluschke et al., who conclude that "in general, there is at present no evidence of large sources of untapped secondary material"[37].

Summary of the Chapter

Unlike many primary commodities, copper is not mined predominantly in developing countries, but also in the "North", especially in North America and the Soviet Union, which in 1982 accounted together for almost 40% of the total world's primary copper output.

Since the second World War, the market power of transnational corporations has declined, and state-owned enterprises have increasingly become active. In spite of the decreasing concentration in mine production and capacities, some barriers to entry remain. Apart from lacking managerial know-how, they are due to the riskiness of explorations and due to immense capital costs for copper mine expansions because of scale economies.

The smelting and refining of copper is even more concentrated in the industrialized economies. Copper and copper alloy scrap is almost only recycled in these countries. The secondary production of the metal differs from primary production as far as costs, especially energy requirements, and prices are concerned. Competitive and highly volatile scrap markets impede collusion in the primary industry.

2.6 Notes

1. The standard reference about the extractive metallurgy of copper is Biswas, Davenport (1976). Very easily readable, but not up-to-date, are the short summaries of mining, smelting and refining by Milliken (1966) and Prior (1966). The following survey is compiled from all the sources mentioned.

2. The influence of smelting on the environment has been one of the main concerns of the copper industry and environmentalists recently. It is, of course, not a new problem. Already more than one hundred years ago, Merbach (1881) reported about severe air pollution arising from copper smelting in the Kingdom of Saxony.

3. The percentages refer to the United States in 1981. See Minerals Yearbook (1981: 289).

4. Biswas and Davenport (1976: 217-241) explain single-step and multi-step processes to produce blister copper.

5. Loc. cit.: 345-346.

6. United Nations (1981) for an analysis of the role and activities of transnational corporations in the copper industry.

7. Piesch (1975) excellently surveys this and other concentration measures. Although Piesch (1982) and Deffaa (1982) develop indices which are more qualified for oligopolistic markets, I chose the concentration ratio due to its simplicity, clearness and vividness.

8. See United Nations (1981: 29-32). The last point is stressed by Labys (1982).

9. This and the following examples are compiled from various issues of the Minerals Yearbook, from trade journals, and from annual reports of copper mining companies.

10. Gluschke et al. (1979: 66-69) describe the problems of measuring capacity.

11. Most figures in this section are calculated from unpublished data, compiled by the Phelps Dodge Corporation Controller's Department.

12. In a study for the World Bank the Charles River Associates (1979) found that three years elapse on an average from the official start of production to the date, when output is finally in line with the designed capacity.

13. See Joralemon (1973: 371) and Minerals Yearbook (1972: 484; 1974: 506).

14. This figure was confirmed to me by industry officials.

15. See Gluschke et al. (1979: 67).

16. The problems of financing mining investments are rather complex and a description of them lies beyond the scope of this book. For an in-depth study of foreign copper mine investments in Papua New Guinea and in Peru see Mikesell (1975). Radetzki and Zorn (1979) analyze the financing of mining projects in developing countries. Their study is summarized in Radetzki (1980) and Radetzki, Zorn (1980). The following account mainly relies on these sources.

17. New York Times, August 13th, 1981, News of the Week, p.5. McGill (1983) studies the financing of the Ok Tedi project.

18. Petrick et al. (1973) analyze the economics of copper's by-product metals.

19. See Minerals Yearbook (1976 :492).

20. Calculated from Mineral Commodity Summaries (1984: 37).

21. CODELCO (1982: 10).

22. See e.g. Mackenzie (1979: 22).

23. Section 4.1 describes the trade relations in the world copper industry.

24. New York Times, December 18th, 1980, News of the Week, p.5. Sousa (1981: 52-56) lists comparative regulatory cost estimates for the United States.

25. It has often been recommended to copper producing developing countries to expand their domestic processing of ores to increase domestic value added and employment (see e.g. United Nations (1972)). This recommendation has been critically evaluated e.g. by Habenicht (1977), Adams, Behrman (1981) and McKern (1981). Gueronik (1974) concentrates on CIPEC's potential for semi-fabricating.

26. Spendlove (1961, 1969) and Davis (1975) survey the methods of secondary copper production.

27. Gordon, Lambo and Schenck (1972b) review the literature on the collection of copper scrap. Bonczar and Tilton (1975) survey important determinants affecting scrap supply and demand.

28. In the next few decades the recovery of copper from electronic scrap (e.g. from multiple-pin plugs and collectors) may gain in importance. See e.g. Salisbury et al. (1981).

29. Calculated from Fairchild Publications (1981: 86).

30. Estimates of the costs of converting scrap to refined copper vary considerably. See Gluschke et al. (1979: 89-91).

31. For data on energy requirements to recycle scrap see Chapman (1974), Arthur D. Little, Inc. (1978) or Kusik, Kenahan (1978), for instance. Data about the U.S. primary copper industry's energy consumption are compiled in Rosenkranz (1976). The lower bound of the estimates in the text above is quoted by Siebert (1979: 411).

32. Section 4.3 describes the structure of the North American copper market.

33. The working of the London Metal Exchange is explained in Chapter 4.3.

34. See Labys, Rees and Elliott (1971) for empirical evidence supporting this hypothesis.

35. The preliminary 1983 figures are calculated from World Metal Statistics, March 1984.

36. Grace (1978) compares recycling efforts in different countries. For West Germany alone see e.g. Grund (1981), for the United States consult Bonczar and Tilton (1975), for instance.

37. See Gluschke et al. (1979: 98). Other studies for the United States point in a similar direction, for example Bergsten (1978: 149).

Chapter 3: Consumption

3.1 Basic Facts

Copper is rarely used as a final product. Generally, the demand for the metal is derived demand.

Immediate consumers are mainly wire rod mills (they used 72% of the total U.S. consumption in 1981), brass mills (26%), foundries (0.6%) and secondary smelters (0.3%)[1]. Only a small amount of refined copper is used to produce chemicals or copper alloy castings, most of it is consumed to fabricate semis, i.e. copper and copper-alloy semi-manufactures. Semis, then, are used im many sectors of the economy.

The main copper consumers are:
- the electrical and electronic products industry (e.g. for telecommunications, power utilities or for wiring devices),
- the construction industry (especially for plumbing, heating, air conditioning or building wiring),
- the transportation industry (with the autobile sector above all),
- the industrial machinery industry (e.g. for in-plant equipment, heat exchangers etc.), and
- the consumer goods and other industries (e.g. for consumer electronics, electric household articles; also for military ordnance)[2].

Of 100 kilograms copper used in the United States in 1980, 28 kg were consumed in the electrical and electronics industry, 30 kg were used for construction purposes, 10 kg in the transport industry and 18 kg for industrial machines. The rest was utilized for durable consumer goods and further articles[3]. In other industrialized countries, the distribution on sectors is similar. Generally, the electrical and construction industries are the main users of copper[4].

Evidently, the use of copper is widely spread over the economy. This is due to the metal's versatile characteristics, including high conductivity (electrical and thermal), tensile strength, corrosion resistance, malleability and, last but not least, its pleasant color[5].

The following sections compile more detailed information about the consumption of copper and about the substitution possibilities.

3.2 Consumption

The total world's copper consumption (direct use of scrap excluded) summed up to 9.1 million tons in 1982 (1972: 7.9 million tons). The market economies' share reached 75% (1972: 79%), almost one third of that was consumed in the European Community.

Industrial countries are the main consumers of copper, as shown by table 3-1. The United States' and the United Kingdom's share declined considerably, all other major consumers' shares increased from 1950 to 1980. Notably Japan's proportion grew more than sixfold in this period.

Table 3-1:

Shares in the total world's refined copper consumption,
in per cent

Country	1950	1960	1970	1980	1982
United States	42.9	25.9	25.5	19.8	18.3
Soviet Union	10.3	13.8	13.0	13.8	14.8
Japan	2.1	6.4	11.3	12.3	13.7
West Germany	6.0	10.9	9.6	7.9	8.1
France	4.1	5.0	4.5	4.6	4.5
United Kingdom	13.6	11.8	7.6	4.3	3.9
Italy	2.4	3.9	3.8	4.1	3.8

In 1980, some 3% of the world's refined production was consumed in the People's Republic of China, in Belgium/Luxemburg, and in Brazil (respectively), about 2% in Canada and Poland (in each), and more than 1% in Yugoslavia, Spain, Mexico, Australia, East Germany and Sweden (respectively). The cumulated amount of refined copper used in the four most important copper exporting developing countries reached less than 1%.

Concentration

Like the concentration of copper production, the concentration of copper consumption declined in the market economies. This development is due mainly to the sharp reduction in the United States' share. Table 3-2 gives some evidence.

Table 3-2:

Concentration indexes for the market economies' refined copper consumption,
in per cent

Year	C(1)	C(3)	C(6)
1950	49	72	83
1960	32	60	79
1970	32	58	78
1980	25	54	71
1983	25	54	70

Total Use of Copper

In chapter 2 it was mentioned that copper and copper alloy scrap is not used
only as input for the production of refined copper. It is also consumed directly,
i.e. used immediately for the production of semis. Therefore, it is worthwile to
look at the total use of copper, i.e. the consumption of refined copper plus the
direct use of copper and copper alloy scrap. These figures thus comprise the
consumption of copper from newly mined ores and from scrap, showing the total
consumption of copper in all forms.

In 1982, the total use of copper in the market economies amounted to 9.1
million tons, only 540,000 tons more than ten years before[6].

Table 3-3 shows some countries' shares in the market economies' total copper
use. Mutadis mutandis, the percentages are similar to the shares in table 3-1, which
referred to the total world. Variations indicate differences in recycling behavior
and international trade in scrap, although relatively small.

Table 3-3:

Shares in the market economies' total copper use,
in per cent

Country	1960	1970	1980	1982
United States	35.2	32.8	28.3	25.9
Japan	9.4	14.8	16.2	18.9
West Germany	12.2	10.7	10.6	10.5
Italy	4.6	5.6	6.3	5.8
United Kingdom	13.8	8.4	5.5	5.3
France	6.4	5.8	5.8	5.7
Belgium/Luxemburg	2.3	2.3	3.5	3.4
Canada	4.7	3.4	2.4	1.7

The total use of copper is concentrated in the industrialized countries, even
more than the copper consumption (excluding the direct use of scrap), because
developing countries use copper and copper alloy scrap only in very small quantities
directly. Scrap data for centrally planned economies is not available.

Concentration ratios can be easily calculated from table 3-3. Not surprisingly, they indicate that the time trend in the concentration of the total copper use also declines.

In summary, this section showed that contrary to copper production, consumption of the red metal is almost only concentrated in the industrialized economies. In these countries, copper is used in many sectors, with the main focus by the construction and electrical industries. Traditionally, this branch has been the sector of the economy, where copper has been subjected most to the "threat" of substitution.

3.3 Substitution

Basic Facts

The replacement of copper by another material, the most important type of substitution, began with the dawn of the Iron Age. It was not before the middle of the 20th century that other materials increasingly did make inroads into copper's preserve[7].

Generally speaking, machines using copper as an input can be used in production with substitutes only with a loss in efficiency, if it is possible at all. For example, a rolling mill producing rod cannot switch easily, and without adjustment costs, to the application of aluminum.

Changes in technology concerning a basic input involve long-term investment decisions and are usually irreversible because of high adjustment costs.

Although long-term substitution is more important for copper users, the short-run substitution possibilities should not be neglected. If in construction, for instance, cold water drainage pipes are used, it is not very important whether they consist of copper, plastics or cast iron. Also, in strip and sheet production aluminum can rather easily replace copper. Thus, in the short-rund substitution possibilities for copper also exist, and here the changes in materials are reversible.

Technological considerations and economic conditions influence the decisions of copper users whether to displace the metal by another material or not.

The technical properties of copper[8], which are important in comparison to its competitors, include its

> - specific gravity,
> - electrical and thermal conductivity,
> - tensile strength,
> - corrosion resistance.

The economic factors comprise

> - absolute and relative prices,
> - the volatility of the market,
> - the availability and apparent abundance of the materials.

The remainder of this chapter analyzes these determinants of copper substitution in detail[9].

Technical Determinants

Copper's main competitors are aluminum, plastics, stainless steel, and, increasingly, fiber optics. Other non-ferrous metals, especially lead, zinc and titanium can also serve in some of the same end-uses as copper[10].

Table 3-4 compares the most important physical and mechanical properties of some of these materials.

Table 3-4:

Technical properties of copper and its substitutes

Material	Specific gravity (g/cm^3)	Relative electrical conductivity[11]	Relative heat-resistance[11]	Tensile strength (kg/cm^2)
Copper	8.96	100	100	630
Aluminum	2.70	60	101	630
Steel	ca. 7.9	2	103-128	880-5,000
Lead	11.36	7	100	6
Zinc	7.13	28	95	210
Titanium	4.51	5	100	1,050

The main technical advantage of aluminum compared to copper is its lightness: one unit of volume aluminum weighs only one third comparatively. Although its electrical conductivity amounts to only two thirds of copper's, there is an edge for aluminum as far as the capacity per unit weight is concerned. Both metals exhibit an excellent workability and about the same heat conductivity. The vibration strength of copper is somewhat superior, also jointing is easier with copper.

Because of these similarities, it is not surprising that the electrical industry is a field of serious competition between aluminum and copper, particularly in electrical conductor applications. In high voltage transmission lines aluminum coaxial cables have completely displaced copper cables. Also, copper sustained losses in the fabrication of low and medium voltage underground cables. Additionally, substitution takes place in heat exchanger applications, e.g. in automotive radiators, air conditioning and refrigerating.

Stainless steel comprises several types of high-alloy steels, whose properties depend on the steel's composition and heat treatment, but which have in common more than 10.5% chromium as an alloying element. In its electrical conductivity copper is far superior to stainless steel. However, whenever high resistance to corrosion, hardness, heat-resistance, ductility and tensile strength are important, stainless steel is preferable to copper. Therefore, between these two metals, substitution is not restricted only to a few fields, as with aluminum, but includes a broad spectrum of possible applications. **Titanium** has similar properties as stainless steels, but it is considerably lighter. Its potential end-uses and competitive applications are alike.

To a small extent, **lead** and **zinc** can sometimes be used as copper substitutes in chemical uses. In addition, a new zinc penny could replace the copper penny in the United States[12] and lead is still used in the building industry for plumbing.

In the latter field, **plastic materials**, however, are copper's most important competitors, especially in tubing and pipes. One of the most important types of plastic, polyvinylchloride, begins to soften above 70°C. Therefore this material is a competitor for copper in coldwater plumbing systems only. Nylon blades have been used for ship propellers, because of their greater resistance to abrasion compared with bronze.

Fiber optics, finally, will probably play a considerable role in the future of the telecommunications industry.

Economic Determinants

Suitable physical and mechanical properties are necessary, but of course not sufficient conditions for the substitution of copper by another material. A comparison of the economic performance of the metal and its competitors is as important.

Prices

From 1960 to 1980 the price of one ton of copper averaged $ 1,709, compared to
$ 957 for aluminum, $ 480 for lead, $ 575 for zinc and $ 5,281 for titanium (in 1975
U.S. dollars[13]). Per unit weight, the prices for stainless steels were in the same
range as the copper price, the prices for plastics were generally somewhat higher.
Per unit of volume, the prices of copper, stainless steels and plastics were roughly
in the same brackets.

As far as the price of copper depends on the energy use necessary to produce
one ton of the metal, copper possesses an advantage compared to its main
competitors. The energy requirements per ton of material amount to
  - 10,000 kWh - 13,000 kWh for steel,
  - 12,000 kWh for high grade copper,
  - 14,000 kWh for copper from 1% porphyry ore,
  - 15,000 kWh for plastic materials,
  - 27,300 kWh for copper from 0.3% porphyry ore, and
  - 56,000 kWh for aluminum ingots from bauxite[14].
Thus, copper production is considerably less energy intensive than aluminum
production, and, depending on the ore from which it is produced, its energy
requirements are similar to those of steel and plastics.

In Napoleon's times, aluminum was more expensive than gold. Gradually, the
price of aluminum in relation to copper declined. In the last twenty years, however,
there exists no significant trend in their price ratio[15]. Fitting time trends for
the price of copper relative to the prices of aluminum, stainless steel, plastics
(polyvinylchloride), and titanium also indicates no time trend in the relative
prices from 1960 to 1980. There was, however, a declining trend of the price of
copper relative to the prices of lead and zinc.

The results for the linear and log-linear time trend are summarized in table
3-5. Fitting linear-log and log-log trends did not change the outcome. The figures
suggest that inroads into copper's application fields by its major substitution
products cannot be explained by changes in their relative prices in the last few
decades.
Indeed, interviews with copper industry officials indicated that the other
above-mentioned economic factors play a more important role in decisions about
long-term substitution. Thus, the volatility of the markets and the availability and
apparent abundance of copper's main competitors are next discussed.

Table 3-5:

Trend of copper prices relative to the price of its main substitutes,
linear and log-linear time-trend[16], 1960 - 1980

Price of copper relative to the price of	Linear Form			Log-linear Form		
	Inter-cept	Slope	R^2	Inter-cept	Slope	R^2
aluminum	27.8 (.627)	-.0132 (-.587)	.018	-1.76 (-.0711)	.00117 (.0932)	.0005
steel	232 (1.04)	-.113 (-.997)	.050	17.7 (0.724)	-.00786 (-.635)	.028
plastics	-173.2 (-.602)	.0932 (.638)	.021	-31.2 (-1.08)	.0170 (1.16)	.066
titanium	1106 (.521)	-5.23 (-.486)	.012	-3.22 (-.111)	.0050 (.0336)	.006
lead	204* (3.13)	-.102* (-3.07)	.332	48.1* (2.23)	-.0238* (-2.17)	.199
zinc	160* (2.17)	-0.008* (-2.12)	.192	54.0* (2.34)	-.0269* (-2.29)	.216

Price volatility

Non-ferrous metal markets notoriously exhibit a high price volatility[17]. Copper
is no exception. Table 3-6 compares the coefficients of variation of current and
1975 prices[18] for the red metal and its main substitutes.

If current prices are considered, the market instability appears to be higher
for copper's substitutes. However, more importantly, figures in constant 1975 U.S.
dollars show that the volatility of the world copper market surpasses the
fluctuations in the markets of all its substitution products.

The reason for this difference is the competitiveness of the copper market
(outside North-America) in contrast to the markets of its main substitutes. These
markets usually are dominated by a relatively few price-setting oligopolists, who,
in pursuit of long-run profit maximization, try to avoid excessive price
fluctuations, because they could induce their consumers to switch to alternate
inputs in the long run, a procedure which is normally irreversible due to the high
adjustment costs.

Table 3-6:

Coefficients of price variation, 1960 - 1980,
in per cent

Material	Country	Coefficient of variation	
		Current U.S.$	1975 U.S.$
Copper	United Kingdom	35.0	29.6
Copper	United States	39.3	15.0
Aluminum	United Kingdom	50.1	10.7
Aluminum	United States	40.4	8.2
Vinyl	United States	35.6	19.4
Steel	United States	45.7	6.9
Titanium	United States	57.0	15.9
Lead	United Kingdom	76.1	28.1

Apparent Abundance

The third possible reason for the substitution between copper and its competitors is the availabiliy and the apparent abundance of the respective material.

In 1980, the reserve base for copper was estimated to be 493 million tons, compared to 95 billion tons of recoverable iron, and 5 billion tons of aluminum equivalent in bauxite.

Total resources (reserves plus subeconomic plus undiscovered deposits) amount to 2.32 billion tons of copper, 40 to 50 billion tons of bauxite and 236 billion tons of iron (always in recoverable metal content)[19]. Potential aluminum resources in non-bauxite materials, although not yet economically exploited, are far greater than the figure mentioned. Indeed, aluminum is the most abundant metal on earth. All common rocks, except lime- and sandstone, contain it as an important component. More than 8% of the earth's crust consists of this metal. With 5%, iron takes the second place in the ranking list of metallic elements, and titanium is already number four. Still 0.4% of the earth's crust is a composite of this metal. Compared with this apparent abundance of its substitution products, copper's share of only 0.007% appears to be very small[20]. It is plausible, therefore, that in the very long run, in an "age of substitutability" (Goeller and Weinberg (1976, 1978)), aluminum, titanium and steel could almost totally replace copper.

Summarizing

Although the price of copper in relation to the prices of its main substitutes did not change significantly in recent years, the higher volatility of the copper prices and the apparent abundance of its main competitors encouraged substitution.

The latter factors are chiefly effective in the very long run, therefore this econometric model does not capture them explictly. Neither measures of price variability nor indices for the availability of the metal prove to be statistically significant determinants of current copper use. But the estimation results demonstrate that, apart from the relative price of copper itself, the relative price of its main competitor aluminum always clearly influences copper consumption, even in the short run. Econometric analysis, as reflected in this model, also indicates that predominantly copper consumption depends on the general level of economic activity. This fact is, of course, not surprising. As stated in the first section of this chapter, many sectors of the economy take advantage of the red metal's versatile properties.

3.4 Notes

1. Calculated from Minerals Yearbook (1981: 296).

2. Smith (1966) surveys typical uses of copper and its alloys.

3. Calculated from Copper Development Association (1981).

4. See, for instance, Gluschke et al. (1979: 29-36).

5. Budinski (1979: 306-322) describes the general properties of copper and its main alloys.

6. Data on the direct use of scrap in the centrally planned economies is not available.

7. For a survey on the substitution of copper and competing metals see e.g. Prain (1980: 151-168), UNCTAD (1977c) or Gluschke et al. (1979: 36-43).

8. See Smart (1954) for a detailed survey of the physical properties of copper.

9. The basic sources for the following information are Flinn, Trojan (1975), Budinski (1979), the standard literature on copper quoted in Chapter 1 and interviews with industry officials.

10. According to Prain (1980: 154-155) "of the losses which copper sustained to alternative materials, aluminium accounted for 54 per cent, plastics 8 per cent, stainless steel 5 per cent and other ferrous metals 18 per cent; the remaining 15 per cent was attributed to design changes, sometimes involving the complete elimination of copper."

11. Copper = 100.

12. The U.S. Treasury estimated that making pennies out of zinc could save the U.S. government $50 million a year (New York Times, April 1st, 1981, News of the Week, p.5). The U.S. Mint Bureau calculated the savings to be $25 million a year (Wall Street Journal, July 26, 1981, p.28).

13. London prices in 1975 U.S.$. Exchange rate and U.S. wholesale price index are compiled from International Financial Statistics. The price of titanium is the average U.S. sponge price published in Fairchild Publications (1981: 223).

14. Cf. Goeller (1979: 153) and Radetzki (1979: 355).

15. Also from 1950 to 1980 no trend can be established. This means that Mezger's assertion of an increasing difference between copper and aluminum prices is wrong (See Mezger (1980: 35)).

16. Asymptotic t-statistics are given in parantheses. A "*" means significant at the 5% level.

17. See e.g. Tilton (1981) or Wagenhals (1981b) for a survey.

18. In 1975 U.S. dollars. Deflator is the U.S. BLS wholesale price index.

19. Data on reserves and resources are compiled by the U.S. geological survey. For the above-mentioned data see Mineral Commodity Summaries (1981: 7, 17, 41 and 77). Chapter 5 describes copper reserves and resources in more detail.

20. For data on the composition of the earth's crust see e.g. Fairchild Publications (1981: 276).

Chapter 4: Trade and Prices

After summarizing some basic information about the general pattern of trade flow in the world copper industry, the first section of this chapter describes how the world copper industry falls into three more or less self-sufficient regions, a fact reflected in the structure of Part II's econometric model. Section 2 reports about the organization of the most important copper exporting developing countries and their efforts to influence prices, while the last section actually explains how copper prices are formed.

4.1 Trade

Basic Facts

Trade in copper is generally trade in the refined metal. Ocean freight rates are still relatively small compared to the value of copper, therefore large quantities of concentrates are traded. Their relative share even has been increasing in recent years. Trade in blister copper is smaller and has been declining[1].

From 1970 to 1980, 59% of the total world exports were traded in refined form, 23% as concentrates and 18% as blister copper[2]. In this decade the total refined exports increased from 2.3 million tons to 3.0 million tons. Table 4-1 indicates a strongly increasing share of the exports of ores and concentrates, which is mainly due to the opening of new pits in some developing countries without smelting facilities.

Table 4-1:

Shares in total world copper exports,
in per cent

Year	Copper ores and concentrates	Blister copper	Refined copper
1970	14.6	21.4	64.0
1980	26.2	14.4	59.4
1982	30.6	16.4	53.0

The next few sections cover in more detail the copper trade within the market economies and between the East and the West.

Exports

The world's most important copper exporters are CIPEC's "Big Four": Chile, Peru, Zaire and Zambia[3]. The Latin American countries' and Zambia's refined copper exports increased considerably in recent years due to the installing of new refining facilities. Today, they export most of their copper in refined form. Only Zaire sells more than 50% of her total copper exports in the form of blister copper.

Papua New Guinea and Indonesia, which began their copper mine production in the early 1970s, play only a minor role in the world trade up to now. They send most of their concentrates to Japan.

Apart from these producer countries, which are members of CIPEC, Canada, the Republic of South Africa and the Philippines are also among the more important copper exporters. Canada and the Republic of South Africa export their copper surplus only, and they consume a considerable amount of their output domestically. The Phillipines, on the other hand, export all their copper in the form of concentrates, mostly to Japan.

The major copper exporters are listed in Table 4-2, which presents WBMS data on copper exports in form of ores and concentrates ("Conc."), blister copper and refined copper in 1970 and 1982. Reliable information about unrefined copper trade in 1950 and 1960 is not available.

Table 4-2:

Copper exports of main producers, 1970 and 1982,
in 1000 tons

Country	1970 Conc.	Blister	Refined	1982 Conc.	Blister	Refined
Canada	162.5	–	265.3	252.8	–	232.6
Chile	38.7	190.1	440.0	200.9	198.9	808.5
Indonesia	–	–	–	76.9	–	–
Papua New Guinea	–	–	–	173.3	–	–
Peru	41.1	135.3	32.6	38.4	97.0	204.1
Philippines	160.3	–	–	280.0	–	–
South Africa	–	102.0	27.9	14.5	83.5	66.4
United States	55.8	7.1	201.4	195.3	2.0	31.9
Zambia	–	103.7	578.4	–	–	602.6
Zaire	–	189.7	180.0	36.0	323.3	156.0
Total exports	534.7	784.3	2348.1	1569.8	842.8	2718.1

Imports

The main trading partners of the copper producers are the Western European countries and Japan. The United States, although also a copper net importer, in contrast to these countries, is almost self-sufficient.

Compared to her total copper useage, Japan's own mine output is small. In 1982 more than 3.6 billion tons of concentrates (gross weight) were imported, mainly from developing countries. Compared with virtually no imports in 1950, this reflects an immense growth in Japan's smelting and refining capacities, so that today, Japan even exports refined copper.

In Western Europe, the Federal Republic of Germany is the largest net importer of copper, followed by France, Italy, and the United Kingdom. West Germany possesses significant smelting and refining capacities (e.g. Hüttenwerke Kayser and Norddeutsche Affinerie). Belgium and the United Kingdom also refine large amounts of copper domestically. Table 4-3 shows the net imports of the major West European copper users. Table 4-4 presents their distribution in the various forms, in which copper is traded.

Table 4-3:

Net copper imports of major European consumers, 1970, 1980 and 1982, in 1000 tons

Country	1970	1980	1982
France	320.3	415.1	381.0
West Germany	521.7	547.2	564.4
Italy	268.5	374.9	319.7
United Kingdom	405.2	313.0	306.1

Table 4-4:

Composition of West European net copper imports in 1982, in per cent

Country	Percentage net imports in the form of ...		
	concentrates	blister copper	refined copper
France	–	5.0	95.0
West Germany	26.6	11.1	62.3
Italy	–	0.3	99.7
United Kingdom	–	22.2	77.8

The ratio of net imports to the total use of copper varied between some 50% for the Federal Republic and about 75% for France in 1980. Because none of the European Community's members mines significant amounts of copper, the different proportions mainly reflect a diverse recycling behavior in these countries.

In the United States, the share of the copper net imports in the total use of the metal averaged only 8% from 1960 to 1980. This is a very small amount compared to the corresponding figures for Japan and the West European countries. Moreover, in these two decades the net imports' share exceeded 14% only twice: in 1967 and 1980, the years characterized by the longest strikes in the United States after the second World War. Thus, the United States is fairly self-sufficient in her copper supply. This fact allows the analytical distinction between an U.S. market and a market consisting of the rest of the market economies. Section 4.3 shows, that the relative seclusion of the U.S. market has important consequences for the determination of copper prices.

Between the United States and the centrally planned economies, virtually no copper trade exists. Nevertheless, in the world copper industry the East-West trade cannot be neglected. It, therefore, will be examined more closely now.

East-West Trade

Copper trade between market economies and centrally planned countries traditionally has been relatively small. Only trade in refined copper is worth mentioning.

Table 4-5 gives some evidence about the East-West trade in refined copper after the second World War. Until the early 1970s the Western countries generally were net exporters, but in the last few years they have become net importers. The ratio of the amount of refined copper net traded between the East and the West to the total use of copper in all market economies together (denoted "import/ use ratio" in table 4-5) has never reached 2% in the last thirty years.

Table 4-5:

East-West trade in refined copper,
in 1000 tons

Year	Western imports	Western exports	Western net imports	Import/use ratio
1950	0.7	12.4	-11.7	-0.4
1960	1.7	96.7	-95.0	-1.8
1970	36.6	96.6	-60.0	-0.8
1980	140.3	99.4	+40.9	+0.4
1982	161.8	120.9	+40.9	+0.4

In the last decade, the trade pattern outlined above has been somewhat breaking up. The main reason is the strongly increasing trade between Poland and West Germany, which has always been the major importer of refined copper from the centrally planned economies. But also growing exports of CIPEC's Big Four to the People's Republic of China have contributed to a change in the copper trade flow in recent years.

To sum up: for the analytical purposes of Part II's econometric model it is justified to partition the entire world copper industry in a non-market segment, consisting of the centrally planned economies, and in a "world market", comprising the Western economies. This "world market" again falls into two more or less self-sufficient regions, the United States and the "rest" of the market economies.

The next section will show that among this remainder some copper exporting developing economies deserve special attention because of their actual or potential attempts to exert their common market power on the working of price formation processes.

4.2 The Council of Copper Exporting Countries

CIPEC, the Council of Copper Exporting Countries, is an organization of the most important copper exporting developing economies.

Its full members are Chile, Indonesia, Peru, Zaire and Zambia. Australia, Papua New Guinea and Yugoslavia are associate members.

In 1982, CIPEC's full members produced 44%, smelted 40%, and refined 26% of the market economies' copper output and they accounted for more than 43% of the Western world's reserves[4]. In regard to world copper trade, they handled 22% of the trade in ores and concentrates, 73% of the blister copper and 65% of the refined copper trade.

Since copper consumption mainly concentrates on the industrialized countries, these figures dramatically indicate the high interdependence on the world copper market and CIPEC's importance as a producer organization.

Therefore it is worthwile, to summarize some information about this organization, its structure, objectives, historical development and its capability of cartel actions. But before this, some more general observations will highlight the interdependence of copper exporters and importers. They will show how CIPEC is firmly interwoven into this fragile balance of importers and exporters.

Interdependence in the World Copper Market

Looking at table 4-6, it is noted that some European countries, and also Japan, depend heavily on imports of blister and refined copper originating in CIPEC countries. Only Belgium and the United States reduced the share of their imports from these countries in recent years.

Table 4-6:

Dependence on CIPEC countries

Country	Import share from CIPEC countries[5]	
	1970	1980
Belgium	91.7	74.2
France	78.2	85.0
West Germany	65.4	68.2
Italy	77.6	88.0
Japan	82.4	83.9
United Kingdom	62.0	58.2
United States	60.6	48.9

On the other hand, the CIPEC countries' exports of blister and refined copper are mainly shipped to a few European countries and to Japan. In 1981, 51% of CIPEC's exports went to the European countries listed in table 4-6 , and 11% to Japan, compared to 62% and 13% ten years earlier.

These few facts strikingly demonstrate the mutual dependence of the CIPEC countries on one side, Japan and the European Community on the other side[6].

But fundamental differences between CIPEC and its main clients influence the relations between copper exporters and importers:
- the CIPEC countries are poor compared with their important customers,
- the share of revenues from copper exports in the CIPEC countries' total export value is considerably greater than Europe's or Japan's share of copper imports in the import value of all commodities[7].

In 1980, the per capita GNP of Chile, Indonesia, Peru, Zaire and Zambia averaged U.S. $ 530, but the per capita GNP of the Western World's five largest consumers averaged U.S. $ 10,960: more than twenty times as much[8]. Although these figures over-dramatize the differences, they give an impression of the dissimilarities in the standards of living[9].

But the CIPEC countries are not only relatively poor, they also depend on the revenues from copper exports[10]. This is shown in table 4-7, where the data for CIPEC's Big Four is supplemented by figures of two other developing copper exporters.

Table 4-7:

Copper's share in exports revenues, average 1970-1980[11], in per cent

Country	Share of copper export value at toal export value, f.o b.	Average rate of export share decline
Zambia	90.9	1.6
Chile	61.1	3.8
Zaire	50.9	4.2
Papua N. G.	48.9	4.3
Peru	20.6	3.1
Philippines	12.3	5.1

Although the share of the value of copper exports in the total export revenues declined for all these developing countries, for Zambia, the country which depends most on her copper exports, this trend is not significant (at a 10% level). Moreover, her coefficient of share variation was largest with 33%, compared to 31% for the Philippines and some 20% for the other countries listed. Together with a lack of sufficient exploration activity and investments in the last decade, this gives a gloomy picture for Zambia's economic future[12].

The other developing copper producers succeeded more in diversifying away from copper. The next table shows some commodities with growing shares in the total export revenues.

Table 4-8:

Export commodities with higher growth rates than copper exports, 1970-1979

Country	Commodities
Zambia	- none -
Chile	Mineral fuels, wood
Zaire	Cobalt, coffee, diamonds
Papua N.G.[13]	Coffee, cocoa, copra and coconut oil
Peru	Lead, coffee, zinc, silver
Philippines	Coconut products, sugar

For the world's main copper consumers, the share of the copper import value in the total imports value is small. Table 4-9 shows that its average value never surpassed 3% from 1970 to 1980. The share of the copper import value declined significantly in this period, due to growing expenditures for crude petroleum imports.

Japan's status as the developing countries' custom smelter and Belgium's position as a transit country in copper trade show up in relatively high import shares
for these countries.

Table 4-9:

Copper's share in import value[14], average 1970-1980,

Country	Share of copper import value at total import value, c.i.f. (in %)	Average rate of decline of share (in %)
Belgium	2.7	10
France	1.7	9
West Germany	1.9	11
Italy	1.7	9
Japan	2.9	10
United Kingdom	1.6	11
United States	0.9	9

Given the high interdependence in the world copper market and the similiarities
of the main developing copper producers, especially their dependence on copper exports to finance their development plans, it is not surprising that since the early
1960s consultations between the main copper exporters began to aim at a mutal repre-
sentation of interests.

Organization

In June 1967, the governments of Chile, Peru, Zaire (then Democratic Republic
of Congo) and Zambia established a consultative organization, the "Conseil International des Pays Exportateurs de Cuivre", generally referred to as CIPEC[15].

This "Intergovernmental Council" is composed of:
- the Conference of Ministers,
- the Governing Board,
- the Copper Information Bureau, and
- the Executive Committee.

CIPEC's highest ranking organ, the Conference of Ministers, implements and coordinates its general policy. The Governing Board coordinates measures related to
all aspects of copper production, marketing and pricing, and towards cooperation
between member countries. The Copper Information Bureau collects and disseminates
statistical and other information relevant for the organization and its member countries. In 1974, the Executive Committee was added, a permanent group charged with
the day-to-day operations of policy making and coordination as well as with public
relations functions[16].

Objectives

CIPEC's objectives are laid down in Article 2 of the 1967 agreement, updated in 1974. Among the targets of the Council are:
- to coordinate measures to achieve a growth in real earnings from copper exports and a greater price stability,
- to harmonize the member countries' policies and promote their solidarity in regard to the copper industry,
- to improve information and to make it available to its members, and
- to foster the coordination with international organizations having similar interests[17].

Whether, and to which extent, these objectives were achieved, will be adressed subsequently, after CIPEC's historical development has been described. But first, to view this organization in comparison to earlier producer associations, table 4-10 lists the most important attempts to regulate the world copper industry in our century[18].

Table 4-10:

Copper producer associations[19]

Organization	Years	Average output share (%)	Maximum output share (%)
Copper Export Assoc.	1918-24	49.4	56.9 (1924)
Copper Exporters, Inc.	1926-32	57.9	64.7 (1928)
International Cartel	1934-39	35.6	44.0 (1934)
International Cartel & associated firms	1934-39	48.9	52.7 (1937)
CIPEC	1967-current	31.4	36.0 (1967)

Table 4-10 proves that CIPEC's average percentage output share and its maximum percentage output share since its foundation are smaller than the respective shares of any of the more important copper producer associations in the twentieth century[20]. Given that none of these organizations was able to influence prices over a considerable time period[21], the prospects for CIPEC do not look good a priori. The following sections deal with this supposition.

History

In the first few years after the foundation of the Council, the organization's work was mainly characterized by information exchange and by attempts to remove administrative and institutional bottlenecks[22]. Although from 1966 to 1971 the ownership patterns of its copper mining industry changed considerably, CIPEC had but loose coordination functions[23].

Enticed by the apparent success of the Organization of Petroleum Exporting Countries, and perhaps by the psychological success, when Japan complied with CIPEC's plea to stop its dumping of excess refined copper at the LME, CIPEC made its first move. In November 1974, CIPEC announced a restriction of its copper production by 10%, effective December 1st, 1974, for the next six months[24]. But this move did not have the effect on prices that CIPEC had hoped for. On the contrary, the LME prices slipped further. Even a raise of the quota to 15% of the exports[25] in April 1975 did not have an effect on prices which can be attributed uniquely to this measure. From 1974 to 1975 the output of the Western World's non-CIPEC copper producers declined more (!) than CIPEC's output (-7.9% vs. -6.3%)[26]. In 1976 the restrictions were formally abolished[27].

Again in 1977, Peru, Zaire and Zambia agreed to reduce copper output and sales in order to reduce the refined copper surplus[28]. But Chile refused to join its fellow CIPEC members, and instead announced a plan to increase her world market share[29]. CIPEC's African members had considerable bottlenecks in transportation and power supply anyway. Peru increased its production after having based its cuts on capacity instead of production[30]. So CIPEC's last attempt, up to now, to influence the international copper market, was also a failure.

In view of these historical experiences, the organization's achievements will now be compared with its objectives.

Goal Attainment

The first and probably the most important goal mentioned above was the increase of real earnings from copper exports and the reduction of the notorious price volatility. Table 4-11 shows the real earnings from copper exports, the average growth rate of real export earnings and the coefficient of variation of the earnings from 1960 to 1969 and from 1970 to 1979 in 1975 U.S. dollars.

The differences in the real export earnings and in the coefficients of variation are not significant statistically. No time trend can be ascertained for the total period 1960-1979, but an increasing trend for the first ten years and a decreasing trend later. These results are valid for each country separately and for all of them together.

This evidence leads to the conclusion that CIPEC did not achieve its objective to increase the real earnings from copper exports.

Table 4-11:

CIPEC's real copper export earnings[31], in billion 1975 U.S. dollars

Country	Average real earnings		Growth rate of earnings		Coefficient of variation	
	60-69	70-79	60-69	70-79	60-69	70-79
Chile	1.129	1.267	10.6	-3.0	33.7	23.2
Peru	0.315	0.346	10.9	-0.9	40.7	31.3
Zaire	0.620	0.610	2.1	-11.3	29.9	44.4
Zambia	1.098	1.090	12.2	-9.1	41.0	36.9
CIPEC[32]	3.162	3.313	9.0	-6.1	32.6	29.7

Table 4-11 also shows the high volatility of the real copper export earnings, which declined only insignificantly over the years. These fluctuations hindered the continuous economic development of the CIPEC countries. Their macroeconomic repercussions pervaded all sectors of the developing economies and were not restricted to mining[33].

Did CIPEC harmonize the member countries' policy and foster their solidarity? Publicly available evidence suggests that this objective was not well achieved. One may think of the unsuccessful cut-backs, for instance. Also, the ideological differences between member countries are considerable. For example, the regular meeting of the Conference of Ministers in November 1973 was delayed for half a year because Zambia's government refused to deal with the Chilean junta[34].

The Council collects and disseminates information about all aspects of the world copper industry. The data is often confidential and is only disclosed to member countries. However, judged from CIPEC's official publications, presumably the organization does a good job in information collection.

To attain the last objective mentioned above, the fostering of the coordination with international organizations pursuing similar interests, contacts with exporters of other raw materials have been created and extended. For instance, since the mid 1970s, representatives of the International Bauxite Association attend the Conference of Ministers as observers[35].

Possibilities of Cartelization

In view of CIPEC's influence on the world market, one might suspect the possibility of a copper exporters' cartel[36].

However, observing the disparity between CIPEC's objectives and the results achieved, I conjecture, and most students of the world copper market agree, that the likelihood of a cartel-like action seems remote.

To analyze in some detail, why the simplistic equation "CIPEC = OPEC" is false, lies beyond the scope of this book. Ideally, one would like to compare the sum of discounted profits for CIPEC's members under the current market structure and under cartelization, taking into consideration the costs of organizing and maintaining the cartel. Pindyck's (1978a) study goes a long way in this direction, but his model neglects very important institutional features of the world copper market, especially the two-tier price system. Furthermore it does not allow explicitly for capacity expansions, inventory behavior et cetera.

Due to the economic and political relevance of a potential copper cartel, there have been many other attempts to analyze whether collusion in copper could work. An outstanding example is the seminal study by Takeuchi (1972), who concludes that "the CIPEC countries must lose in the long run if they attempt to jack up the price of copper by cutting back their supply on a long-term basis" (loc. cit.: 28). Panayotou (1979a, b), while tentatively also including Canada as a cartel member, obtains similar results[37].

Additionally, many non-quantative contributions address the problem[38]. From a banker's point of view (Citybank (1978)) to an opinion from the New Left (Bourderie (1974)) their conclusions are similar: a CIPEC copper cartel would most likely not be successful in augmenting the real foreign exchange earnings of its members.

Eventually, in the future, with imminent new entrants in the world copper market, with considerable capacity increases planned in non-CIPEC countries, and with the potential competition of copper from seabed mineral resources[39] the potential of the CIPEC countries to control the world copper market will certainly not increase.

4.3 Prices

Basic Facts

The description of the world trade patterns in the first section of this chapter made it clear that the copper world industry can be devided into three rather self – sufficient segments:

- the North-American market,
- the other market economies, and
- the centrally planned countries.

In each of these regions different pricing systems exist.

In the centrally planned economies copper prices are set to achieve political or economic goals without relying on allocation by markets[40].

In the market economies, the main part of concentrates, blister and refined copper is sold on bilateral contracts between producers and consumers. Outside North-America, the price of refined copper is generally based on an average of price quotations at the London Metal Exchange (LME). The prices of concentrates and blister copper allow for smelting and refining costs, for by-product credits or charges and so on. They are very closely related to the competitive LME prices[41].

The North American pricing system is somewhat more complex. For a long time, the main copper producers fixed prices unilaterally, changing them only infrequently, but always in view of the LME price. More recently their prices approximated very closely the quotations of the New York Commodity Exchange (COMEX), a competitive market, whose prices are highly correlated with the corresponding LME quotations.

The market for copper and copper alloy scrap in North America and elsewhere has always been very competitive, and scrap prices tend to be in line with the LME and COMEX prices.

Figure 4-1 shows the development of the U.S. producer price and the LME price (in 1980 constant U.S. dollars) from 1950 to 1983[42]. One glance at the figure shows the periods of the two-tier price system, when U.S. and world market prices differed significantly[43].

The next subsections describe in some more detail the markets where the copper prices are determined.

Figure 4-1:

U.S. producer price ("x") and LME price ("*"), electrolytic copper, wirebars, 1980 constant U.S. $ per ton, cubic spline interpolation

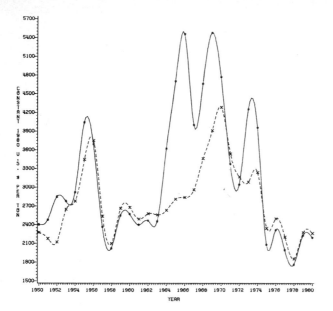

The London Metal Exchange

The London Metal Exchange is the most important market for copper outside North America.

Measured in terms of its copper turnover, the importance of the LME increased from 1970 to 1980. In this period the turnover grew 6% on an annual average to reach 4.5 million tons in 1980[44].

The main functions of the Exchange are:
- to determine a free market price for some types of refined copper,
- to serve the hedging needs of copper market participants, and, to a smaller extent,
- to allow for speculation and for hedging against currency and inflation risks.

Since the first copper contract was introduced in 1883, the LME has always been a sensible, even - in Sies' terminology - "nervous" price barometer reacting quickly on economic and political changes, expected and actual.

Trading at the LME takes place in the "Ring". For five minutes, twice daily, bids and offers meet by open outcry. Patronage or discrimination between ring members are not permitted[45]. Purchasers and sellers act as principals, i.e. they themselves are responsible for financial settlements, in contrast to all other commodity markets, where clearing houses are the rule.

To avoid high transaction costs, copper trading on the LME has always been confined to just a few shapes. In 1963, the Standard Copper contract was replaced by contracts for wirebars, fire-refined and cathode copper. Because of the growing importance of continuous casting in the last few years, the weight shifted gradually from wirebars to cathode copper. Due to this development, in 1982, the CIPEC countries switched from the wirebar to the cathode contract as the basis of their copper pricing[46].

It is important to stress that generally only excess copper is traded on the LME. The Exchange has always been a "market of the last resort for physical copper" (Wolff (1980: 105)). Far more important than LME's cash market is the futures market in copper contracts. This market yields the highest volume in trading. Primarily, the LME is used for hedging purposes. A typical LME transaction is the opening of a futures position and simultaneously taking an opposite position in the actuals market to reduce price risks[47].

Dating back to the time, when copper was shipped from Chile to England in three months, the duration of the official LME futures contract is limited to three this period[48]. In New York, COMEX offers trading in 23 months futures contracts[49]. The market for these long-term contracts is thin, probably because the costs of alternative ways of risk reduction are smaller for such a long period[50].

Besides being a price barometer and offering an opportunity to hedge against price risks, the LME offers a possibility to hold uncovered positions, i.e. to speculate[51]. Between 20% and 30% of its transactions can be attributed to speculation (Wolff (1980: 105)). However, in some periods the share of the speculative element was significantly higher, e.g. in 1973, when the LME's turnover grew 86% compared with 1972[52].

The LME essentially is an international market. Many of its representative subscribers are employed by foreign companies and most of its transactions originate outside the United Kingdom. The U.S. dollar is the most important currency in the international copper trade, but the LME quotes copper in Pound sterling. Thus, there is both a need and limited scope for short-term hedges against the risk of parity changes of these currencies and for speculation. For example, if a British citizen expects a decline of the Pound sterling in respect to the U.S. dollar, he or she can buy copper as a hedge, because by law he or she is not allowed to take a speculative position in dollars[53].

Generally, actual and expected parity changes influence spot and futures LME prices. Depreciations of the Pound usually lead to increases of the LME copper prices, because these prices are determined by the fundamentals affecting the entire world copper market. An important consequence of this countervailing price increase is a mitigation of the factor cost growth of copper using fabricators outside the United Kingdom (Sies, 1981c).

In times of high inflation rates, metals are sometimes considered to be "a refuge of value", "a store of wealth", et cetera. Money is "invested" not only in precious metals, but also in copper due to its high marketability at the LME[54]. Thus, inflationary expectations may bias relative copper prices and distort an optimal allocation of the metal. The most recent example was the rise of the copper futures prices during the 1981 bull market in precious metals.

Producer Prices in North America

In North America, the New York Commodity Exchange (COMEX) offers similar hedging and speculation opportunities as the LME[55].

Arbitrage transactions between the two markets are common and lead to highly correlated prices. At COMEX, the degree of speculation is higher and the volume of transactions is smaller than at the LME. Unlike in London, at COMEX a clearing-house system exists. Nevertheless, the risks for long hedgers wanting to take delivery are higher in New York, because more types of copper are tenderable against a COMEX contract. Grade, delivery point, and delivery date in a trading month are at the seller's discretion only. Also, limits to price changes can create risks for arbitrageurs in bull or bear markets.

Since 1978, the Exchange's copper contract provides the pricing basis for many large copper companies in North America.

Up to the 1970s, the United States' and Canadian copper market can be described as an oligopoly with a competitive fringe. For a long time, the Kennecott Minerals Company has been the largest U.S. producer, followed by the Phelps Dodge Corporation[56]. Both companies frequently were price leaders. In Canada, the mainstay of copper production is the world's major nickel producer, INCO Ltd., followed by the divisons of Noranda Ltd.

Most of the large copper producing companies are vertically integrated from production to the refining stage, some are partially integrated forward into semi-fabrication. Many producers still maintain an important role outside North America[57]. Generally, the output of these companies is not restricted to copper, but comprises many minerals. Since the early 1960s, oil and gas companies have attempted to aquire copper companies[58], the most successful example is the absorption of Kennecott by the Standard Oil Co. (Ohio) in 1981.

Ownership ties and contractual arrangements are rather complex in the North American copper industry. Therefore any evaluation of the degree of concentration and the intensity of competition has to be cautiously considered. But all available evidence suggests a decrease in concentration and an increase in competition between the leading U.S. and Canadian copper producers in the last thirty years.

This opinion is supported by figures 4-2 and 4-3, which show the cumulative percentage shares of the largest copper producers, ranked according to their output, in the total output of the ten leading copper companies in 1960 and 1980, both for the United States and for Canada. They indicate a clear-cut decline in concentration among the ten leading copper producers in both countries.

Figure 4-2:

Cumulated percentage shares of ten leading copper mining companies in their total copper output, United States, 1960 and 1980

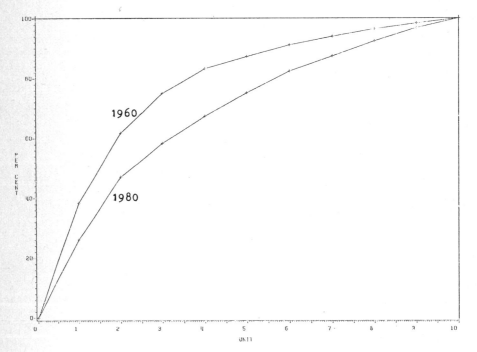

Figure 4-3:

Cumulated percentage shares of ten leading copper mining companies in their total copper output, Canada, 1960 and 1980

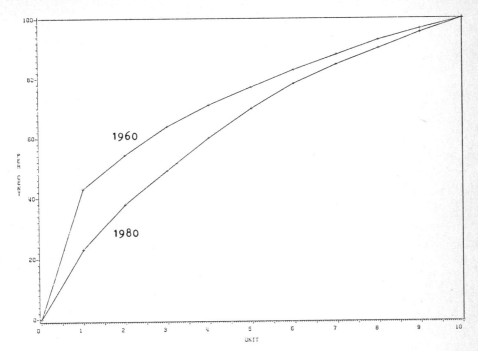

This decline in concentration is one of the reasons of increasing competition in both countries. Another one is the abandonment of export controls, which hampered copper imports from early 1964 to mid 1970 for the United States[59].

The growth of competition is reflected in a change of the North American copper companies' pricing behavior. For a long time the major producers held their prices relatively stable. In times of heavy demand they rather rationed production and lowered their stocks instead of increasing their prices[60]. But, in May 1978, Kennecott abandoned the producer price tradition and most of the large companies followed. They next tied their cathode price to the COMEX price and therefore to the world market price determined at the LME[61].

In March 1982, this approach was changed again. Now, the linkage to the COMEX price is only maintained above a floor price covering variable costs[62]. Phelps Dodge did not follow, but shut down all its copper mines and three of its four smelters for three months to avoid heavy losses due to insufficient prices[63]. Most of the remaining mines curtailed production considerably.

Summarizing, after the second World War the primary copper market in North America gradually developed from a loose oligopoly with two price leaders to a pricing system which is much closer to auction market pricing.

The North American Scrap Market

Contrary to the primary copper market, the scrap market in North America has always been competitive. Although two large custom refiners, ASARCO Inc. and AMAX Copper Inc., use most of the scrap in the United States, the prices of copper scrap and copper alloy scrap are basically driven by market forces. This is, because "scrap is generated, collected, processed and consumed by a very large number and variety of concerns" (Gluschke et al. (1979: 90)).

According to the copper content and to the degree of alloying there are several types of scrap, which were described in section 2.5. Among these types, the price of #1 scrap is highly correlated with the U.S. producer price[64]. In many uses # 1 scrap serves as input in fabricating, without or with only a small amount of processing. The price of #2 scrap varies more in line with the LME futures prices, as the LME price is competitive and because it takes time to collect and process #2 scrap[65].

While consumers of primary copper generally buy most of their copper on long-term contracts requiring them to accept a certain amount within a given range, secondary copper is traded in short-term contracts between dealers, merchants, secondary smelters, foundries, brass mills and other fabricators. This makes scrap prices more liable to fluctuation: from 1956 to 1980 the coefficient of variation of the U.S. producer price was 16% in compared with a 27% coefficient of variation of the #2 copper scrap price[66], a situation typical for recycled materials[67].

Thus, the pricing of copper and copper alloy scrap in North America reflects mainly the conditions of the 'outside market', and is linked to the London Metal Exchange free market price. Strong competition characterizes the secondary market and impedes collusion in the primary copper industry. These features reappear in Part II's econometric model.

4.4 Conclusions

This chapter described the trade pattern in the world copper market and explained the pricing of the metal.

The world copper industry can be divided in a market segment, consisting of two
elements, namely North America and the remainder of the Western countries, and a
non-market segment, consisting of the centrally planned economies.

World copper trading takes place mainly between the market economies outside
North America, especially between Japan and Western Europe on the one hand and the
developing copper exporting countries on the other hand.

The most important copper exporting developing countries, which often highly
depend on their mining sectors to finance their development plans, have formed the
"Council of Copper Exporting Countries" (CIPEC). This organization did not succeed
in increasing the real export earnings of its member countries, its most important
goal. Also, collusion to influence the world copper price did not work.

The relative seclusion between the two segments of the world copper market has,
at times, led to a two-tier pricing system for copper. The LME price served as the
world copper price and producer prices served as domestic prices in the United
States and in Canada.

Since the second World War, the primary and secondary copper market in North
America presents a composite picture of weak and decreasing oligopolistic collusion
between the major vertically integrated primary producers, and of increasing competition in both markets. This picture of an increasing competitiveness in the North
American copper industry matches the framework of the world copper market, where a
few companies' control over a large share in the world's copper reserves immediately
after the second World War eroded to an essentially competitive industry today.

4.5 Notes

1. In this Chapter, the data from Metallgesellschaft's <u>Metal Statistics</u> is supple-
    mented by figures from World Bureau of Metal Statistics (1977) and from various
    issues of <u>World Metal Statistics</u>. Additionally information from trade journals,
    from annual company reports and from the <u>Minerals Yearbook</u> is included. Chapter
    4 in Wagenhals (1983a) presents a more detailed analysis of the world copper
    trade.

2. For a general survey of the trade in non-ferrous metals see e.g. Sies (1981a) or
    Müller-Ohlsen (1981: 167-175).

3. Cf. Dayton (1979). The next section deals with CIPEC in detail.

4. The respective percentages based on the total world output were 33% for the mine, 31% for the smelter, 19% for the refined production, and 38% for CIPEC's share in the total world's reserves.

5. To avoid double counting, only imports from non-European sources are included.

6. An examination of dependence and interdependence in the marketing of copper and in technical cooperation lies beyond the scope of this book. Moran's (1974) and Cunningham's (1981) country studies deal with this problem. Ilunkamba (1980) attempts to describe specifically the "dependence maintained by technology" for Zaire. More references on technical dependence are quoted by Mezger (1980: 62-65 and 78-84).

7. Similar statements are true for copper export revenues in relation to GNP, GDP and government revenues. See e.g. Banks (1974: 5) and Bostock, Harvey (1972: 4).

8. The figures are calculated from World Bank (1982).

9. Preferable would be comparisons of real income measures based e.g. on the United Nations' International Comparisons Project (see Kravis et al. (1975, 1978, 1982)). However, some of the relevant countries are not (yet) covered by these studies.

10. This is not true for Indonesia. She exports primarily crude petroleum, rubber and tropical wood and usually depends on exports of copper concentrates for less than 1% of her total export revenues. From 1975 to 1979 copper's share averaged 0.7%, its coefficient of variation was high with 31%. (Calculated from the United Nations' Yearbook of International Trade Statistics.)

11. Calculated from International Financial Statistics. Papua New Guinea since 1973 only, Peru and Zambia without 1980.

12. The figures for Zaire are somewhat deceiving, because Zaire's second very important export commodity is cobalt, often produced jointly with copper. From 1970 to 1979, Zaire's dependence on cobalt and copper exports together grew from 72% of the total export revenues to 82%, due to a very high cobalt price in 1979, but dropped to 64% in 1980.

13. 1973 - 1979.

14. Calculated from Yearbook of International Trade Statistics, various issues.

15. For general surveys of commodity market agreements see e.g. Radetzki (1970), Law (1975), Behrman (1977, 1978a, 1979) or Nappi (1979). Zorn (1978) analyzes producer associations with special reference to the copper market. For a description of CIPEC see e.g. Gueronik (1970, 1975), Clarfield et al. (1975), Mingst (1976), Dayton (1979), Martner (1979: 89-95) and Nappi (1979: 103-120).

16. Clarfield et al. (1975: 75-78) study CIPEC's formal organization in more detail.

17. See CIPEC (1974).

18. A history of earlier attempts to regulate the copper market is beyond the scope of this book. See e.g. von Broich-Oppert (1926), Skelton (1937), Fink (1948), Cohen (1952: 120-126) or Cordero, Tarring (1960).

19. The market shares are calculated from Herfindahl (1959: 124-125) and from various issues of Metal Statistics.

20. Incidentally, from 1950 to 1980, the average share of CIPEC's foundation members in the total world production was 33.5%. Their average share in the market economies' output amounted to 44.4% in the same period. The CIPEC countries' share in total world production decreased significantly (at a 10% level) in this period. No trend can be ascertained for the share of the non-CIPEC market economies, so that CIPEC's relative decline in the last three decades is due to the growing copper production in the centrally planned economies.

21. Herfindahl (1959: 67-152) and, updating him, Mikesell (1979: 104-106) study the history of copper prices.

22. For evidence cf. Gueronik (1970, 1975) and Mingst (1976: 275).

23. See e.g. Zorn (1978: 225-226). Among the more detailed studies of the (partial) nationalization of the copper industries in the CIPEC countries see Girvan (1970), Moran (1974) and Faundez (1978) for Chile, Potter (1971), Fry, Harvey (1974), Murapa (1976) and Chapter 10 of Cunningham (1981) for Zambia. For Peru and Zaire cf. the country studies in Seidman (1975).

24. _Metal Bulletin_, No. 5943, November 22nd, 1975, p.20.

25. _Metal Bulletin_, No. 5982, April 15th, 1975, p.16.

26. See also Zorn (1978: 228-229).

27. Cf. Wall Street Journal, June 14th, 1976, p.16.

28. Cf. New York Times, March 3rd, 1978, News of the Week, p.5.

29. Cf. Wall Street Journal, March 3rd, 1978, p.20.

30. Cf. Asian Wall Street Journal, July 21st, 1978, p.6.

31. Deflator is the export unit value index of the industrialized countries published in _International Financial Statistics_.

32. Excluding Indonesia, whose share in the world production was only 0.7% in 1980, and Mauretania, whose copper production has ended.

33. A general analysis of these fluctuations and their effects on developing economies cannot be given here. A large amount of research has been dedicated to this topic. For surveys see e.g. Wilson (1977) or Adams, Behrman (1982).

34. _Metal Bulletin_, No. 5847, November 2nd, 1973, p.13.

35. _Metal Bulletin_, No. 6045, November 28th, 1975, p.24-25.

36. It is misleading to characterize CIPEC as a cartel, like Mezger (1980) constantly does.

37. As Professor M. Faber pointed out, it may be very interesting to analyze the potential market power of a copper export cartel with game theoretical methods (see e.g. Selten (1973)). In this respect, Professor H.J. Jaksch's seminal analysis of oligopolistic supply on the world cocoa market (i.e. Jaksch (1982)) is the only example of a synthesis of procedures from econometrics and game theory for commodity markets up to now.

38. See, inter alia, Bergsten (1974), Krasner (1974), Mikdashi (1974), Mikesell (1974), Lindert and Kindleberger (1982: 204).

39. For references to econometric studies, which deal with the impact of ocean floor nodules on the world copper market see Chapter 6.

40. See e.g. Strishkov (1979: 242).

41. Labys (1980a: 98-99) describes the pricing of concentrates and blister copper in more detail.

42. From Commodity Trade and Price Trends, August 1981, p. 98.

43. McNicol (1973, 1975), Felgran (1982) and de Kuijper (1983: 111-199) study the two tier price system.

44. Calculated from Wolff (1975: 199) and Sies (1980a: 154).

45. For a description of the structure and membership of the LME see Wolff (1980: 41-51).

46. See Metall, 35. Jahrgang, Heft 11, November 1981, p. 1154.

47. There are many introductions to the practice of futures trading and especially to hedging, e.g. Sharpe (1978: 389-429). Hedging at the LME is described in some detail in Wolff (1980: 58-93).

48. This temporal limitation has not only traditional reasons. Streit, for instance, remarked that there is no demand for the hedging of excess quantities. See the discussion in Siebert (1980: 495).

49. See Commodity Exchange, Inc. (1982: 7).

50. In a more general setting, this point is suggested by Newbery and Stiglitz (1981: 192) to explain why most futures trading is limited to short-term contracts.

51. Labys, Thomas (1975) very carefully examine the relations between hedging, speculation and the LME prices.

52. Calculated from Wolff (1975: 199).

53. See Mikesell (1979: 93). He notes that this kind of currency speculation probably had an effect on prices in 1976.

54. In view of the 25 tons minimum contract at the LME it is quite unlikely that copper really is the "'poor man's' inflation hedge" as suggested in Standard and Poor's Industry Surveys (1981: M 177). Even for the Italian trade this minimum contract is too large (Nahum (1975)), let alone for any poor man.

55. For the history of COMEX see e.g. Commodity Exchange, Inc. (1977). The following information is compiled from Wolff (1980: 223-226), from Commodity Exchange, Inc. (1982) and from an interview with a COMEX trader.

56. Tomimatsu (1980) and Souza (1981) analyze the corporate structure of the U.S. copper industry. Mikesell (1979: 38-44), Navin (1978), Stewardson (1970: 170-173) and United Nations (1981), among others, include information about copper companies outside the United States. The history of the American copper industry with many references to ownership ties is described in May (1937) or McCarthy (1963, 1964).

57. See Sideri, Johns (1980: 11).

58. For a table showing oil company acquisitions of copper mining companies from 1963 to 1981 see New York Times, April 12th, 1981, p. 3.

59. More evidence on the competitiveness of mineral extracting industries in Canada and the United States is compiled e.g. in Anders et al. (1980: 50-62).

60. The main reason was probably the attempt to discourage substitution. McNicol (1975) offers other motives, but concedes that his arguments are "not conclusive" (p. 50) and "only partially substantiated" (p. 73).

61. See e.g. Wall Street Journal, August 1st, 1978, p. 32.

62. New York Times, March 24th, 1982, p. D-1.

63. New York Times, April 8th, 1982, p. D-6, and Business Week, May 3rd, 1982, p. 80.

64. This assertion is based on a calculation of coefficients of determination for annual prices from 1956 to 1980 (in 1975 U.S. dollars).

65. Gluschke et al. (1979: 89-92) describe the pricing of scrap in more detail.

66. Both prices are deflated with the BLS wholesale price index.

67. See e.g. Dasgupta, Heal (1979: 212).

Chapter 5: Reserves and Resources

This chapter evaluates copper reserves and resources. It assesses the long-run perspectives of availability and analyzes implications for the quantitative modeling of copper industry behavior.

5.1 Copper Deposits

Copper bearing deposits occur in all parts of the world. The copper content of the earth's continental crust is about 70 parts per one million parts of rock (ppm) or almost $3*10^{13}$ tons[1]. Therefore the crustal abundance exceeds the 1980 reserves some 60,000 times.

Although ocean waters contain minute amounts of copper[2], in regard to the total resources of the red metal it is more important that they cover deep sea nodules, which can contain significant quantities of copper.

Above-crustal averages of copper can be found in magmatic rocks (liquid-magmatic and hydrothermal deposits) and in sedimentary deposits (including volcanic-sedimentary deposits).

Porphyry deposits play the most important role in world copper production, followed by sedimentary deposits. Only a small amount of copper is mined from magmatic-contact-metamorphic or metasomatic deposits[3].

Porphyry deposits occur preponderantly in Chile, Peru, Mexico, the United States, Papua New Guinea and the Philippines. Sedimentary and volcanic sedimentary deposits can be found in Central Africa, Australia and Japan. Magmatic-contact-metamorphic and metasomatic deposits occur in parts of Canada and the Soviet Union[4].

Like aluminum and iron, copper follows a "resource triangle", i.e. occurences increase with decreasing concentrations or qualities[5]. Porphyry deposits exhibit lower ore grades than the other types of deposits mentioned. But technical progress and by-product credits make it possible to mine copper economically with a very low grade, currently some 0.2%.

A few centuries ago, only very high grade copper deposits were mined. For instance, from 1771 to 1786 the percentage yield of copper ores mined in Cornwall averaged 12% (Whitney, 1854)[6]. This also was the average ore grade of all Royal Saxonian mines in 1844 (Königliche Bergacademie (1845: 67)).

Jankovic (1967) analyzed 300 copper deposits from 1880 to 1960 and proved the existence of significantly declining time trends of the ore grades. My own analysis for selected mines indicated that this trend continued in the last twenty years. Most specialists agree that the average ore grade mined will deteriorate further[7].

To illustrate these arguments, figure 5-1 shows the development of the average U.S. ore grade since the mid 1920s[8]. The peak in the early 1930s is due to the depression, when mainly high yield / low cost mines were operating[9].

Figure 5-1:

Average copper ore grade, United States, 1924-1982

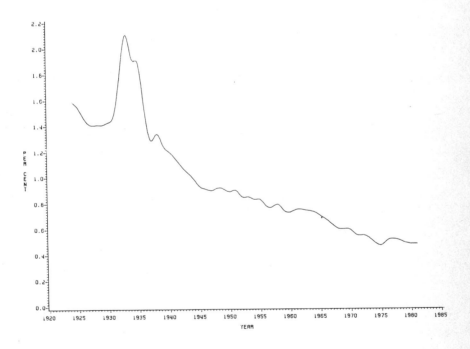

An Appraisal of Copper Resources and Reserves

Definitions of the terms "reserves" and "resources" vary considerably in the literature[10]. Recently the U.S. Geological Survey and the U.S. Bureau of Mines introduced principles of a reserve / resource classification for minerals[11].

Following these definitions, a **copper resource** is a concentration of copper-bearing material in a form and amount, such that profitable extraction is feasible, under defined assumptions about investment. These resources are identified, if their "location, grade, quality, and quantity are known or estimated from specific geologic evidence"[12]. A part of the resource, characterized by meeting certain minimum physical and chemical criteria related to current technology, is called **reserve base**. It consists in **reserves**, i.e. currently economic resources, and in resources that are only marginal or subeconomic. Thus, "reserve base" is a category based on technological criteria, whilst its partition in components depends on economic criteria.

In 1983, the world's copper reserve base amounted to 510 million tons, compared to 493 million tons in 1980 and 279 million tons in 1970[13]. Between 1950 and 1980, the world's copper reserve base rose by a factor 5, while copper consumption rose only by a factor 3.1[14].

Table 5-1 shows the major copper producers' share in the total world's copper reserve base in 1970, 1980 and 1983. Chile and the United States together dispose of almost 40% of all reserves, the rest is widely spread all over the world. Besides the countries listed, worth mentioning are the Philippines (4%), Australia, Papua New Guinea and Poland (3% respectively) and the Republic of South Africa (1%)[15].

Table 5-1:

Distribution of world copper reserve base and reserve/extraction ratio[16]

Country	Percentage share in copper world reserves 1970	1980	1983	Reserve / extraction ratio (1983)
Chile	19.3	19.7	19.0	78
United States	27.8	18.7	17.6	86
Soviet Union	12.5	7.3	7.1	36
Zambia	9.7	6.9	6.7	59
Peru	8.0	6.5	6.3	97
Canada	3.2	6.5	6.3	53
Zaire	6.5	4.9	5.9	61

Table 5-1 also indicates: current copper production can be sustained for a long time, even if the reserve base were fixed at its 1983 level. For each of the most important copper producing countries, the last column in this table shows the "reserve / extraction ratio" defined as the quotient of the copper reserve base in 1983 and of the same years's output. Even under the quite unrealistic assumption that suddenly the copper reserve base would cease to grow, the market economies could extract copper at the current rate for 72 more years on average. This is more than two generations, and therefore longer than the actual planning horizon of any copper company or state mining agency.

A calculation of concentration ratios shows that the concentration of the reserve base on countries decreased. CIPEC's share declined from 44% in 1970 to 38% in 1983. Probably the most important reason for this development was increasing political risk perceived by potential foreign investors, which led to reduced exploration activities in these countries[17].

The world's total copper resources were almost five times greater than the reserves in 1983, and even the land-based resources alone surpassed them more than three times[18]. Excluding resources in ocean floor nodules, 23% of the remaining land-based resources are estimated to exist in the United States, 17% in Chile[19], 14% in centrally planned economies and the rest widely dispersed all over the world.

The information on reserves and resources compiled above is the most reliable data currently available. Yet it is necessary to stress that estimates of reserves and resources in the past increased constantly and varied considerably. Table 5-2 lists various estimates of the world's copper reserve base since 1950 to underscore this point. More clearly, a look at figure 5-2 shows an increase in the reserve base estimates during the last thirty years and a decline in the variance of the estimates in the last few years.

Table 5-2:

World copper reserve base estimates since 1950

Year	Source	Million tons
1950	Tilton (1977a: 10)	100
1950	Paley Commission (1952)[20]	181
1956	U.S. Bureau of Mines	> 90
1960	U.S. Bureau of Mines	154
1964	U.S. Bureau of Mines	140
1965	U.S. Bureau of Mines	192
1966	Schröder (1966: 12)	500
1970	U.S. Bureau of Mines	279
1971	U.S. Bureau of Mines	348

Table 5-2: (continued)

Year	Source	Million tons
1973	U.S. Bureau of Mines	370
1974	U.S. Bureau of Mines	298
1975	U.S. Bureau of Mines	408
1976	Bowen, Gunatilaka (1977: 17)	420
1976	Gluschke et al. (1979: 53), Survey A	451
1976	Gluschke et al. (1979: 53), Survey B	445
1976	Krauss (1976)[22]	456
1976	U.S. Bureau of Mines	459
1977	U.S. Bureau of Mines	458
1978	U.S. Bureau of Mines	549
1979	U.S. Bureau of Mines	498
1980	U.S. Bureau of Mines	493
1981	U.S. Bureau of Mines	505
1982	U.S. Bureau of Mines	511
1983	U.S. Bureau of Mines	510

Figure 5-2:

Estimates of the world's copper reserve base since 1950

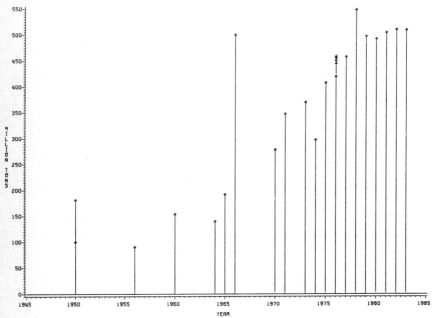

After this assessment of the world's copper reserves and resources, it is quite
natural to ask: "How long will they last?" Therefore, the availability and scarcity
of the red metal is now examined.

5.2 Perspectives of Copper Availability

It is beyond the scope of this book to survey and criticize all measures which are supposed to signal the scarcity of a natural resource[23]. This chapter deals with some of the more common measures, reviews the empirical evidence in regard to copper, and presents some new results concerning the development of the real copper price in terms of various numeraires.

The ultimate geological frontier of copper availability is the crustal abundance. The quantity, which can be recovered, given the current state of knowledge, determines the copper resources; the amount, which can be recovered economically, determines the reserves[24].

Crustal abundance, resources and reserves, these three potential criteria of copper availability can be readily transformed into measures of copper scarcity by relating them to current consumption.

In 1980, the ratio of copper in the continental crust to the annual world consumption was greater than two million, the ratio of land-based resources and of reserves to consumption was 170 and 52 respectively[25].

Crustal abundance, the first scarcity measure, evidently is very optimistic, because copper is widely diluted in the earth's crust and, of course, not every single copper atom in the earth's crust can be recovered.

The other two measures are too pessimistic.

First, resources are defined in terms of a given, known technology. Thus, resources are not fixed once and for all, rather their evolution depends on the amount of effort allocated to research and to the development of new technologies.

Secondly, only if all factors mitigating a potential copper scarcity are neglected, then the reserve / consumption ratio shows in how many years the world will run out of copper at the current rate of consumption. But up to the present day, the history of the copper industry has proved the existence of numerous influences, which have alleviated a potential scarcity. Just a few examples support this claim:

- the implementation of technological innovations: remote sensing from satellites, for instance, has reduced exploration costs, and, e.g. has recently led to the discovery of large copper deposits in remote parts of Pakistan,
- new developments in the efficiency of copper recovery: the utilization of oxide ores has been improved through modern hydrometallurgical processing and in situ leaching techniques,
- increases in the economies of scale: the introduction of the Jackling process early in this century allows the mining of porphyry ore bodies economically,

- improved recycling and all types of substitution, which were considered in the third chapter.

The reserve / consumption measure calculated above is the "static measure" of resource availability in the Club of Rome's first report[26]. If applied to the estimated reserve of 100 million tons of copper in 1950, then copper's doomsday would have been 1983.

According to the Club of Rome's "dynamic measure" of availability[27], which is based on a given positive growth rate of consumption, the world would have run out of copper even earlier.

Such calculations indicate, that the reserve / consumption ratio is not a very reasonable measure of resource availability[28]. Geological and economic adaptation processes are neglected, in contradiction to historic experience. Therefore a reserve / consumption ratio should not be interpreted as a signal of scarcity, but as a warning signal, which shows that adaptation processes are necessary, and which reveals the urgency and extent of the necessary adaptation processes[29].

After the dismissal of the ratio of crustal abundance, resources and reserves to consumption, the next few sections turn to more promising measures of scarcity:

- the rental rate,
- the unit cost of extractive output, and
- the real price of copper.

Rental Rate

Most economists agree that the rental rate is the best measure of scarcity for a resource in situ[30], because it can be interpreted as the opportunity cost due to the use of the resource now[31].

Marginal discovery costs, e.g. prices for licenses granting prospecting rights for potential copper deposits, might be a suitable proxy, but neither such data, nor, of course, figures for the rental rate itself are available.

Therefore, in applied studies, mostly unit cost of extraction indices or relative prices are used to assess the scarcity of a resource.

Unit Costs of Extractive Output

The unit costs of extractive output are defined as the ratio of a weighted average of input of labor and reproducible capital (i.e. excluding the value of the in situ resource) to the output of an extractive industry adjusted to exclude the value of the purchased inputs[32].

Barnett and Morse (1963: 169-170) find that the cost per unit of mineral output declined significantly from 1890 to 1957, no matter whether costs are measured in terms of labor or in terms of labor and capital. For the period after 1957, Johnson, Bennett (1980) and Johnson, Bell, Bennett (1980: 264-265) find evidence of increasing copper scarcity measured by the unit cost trend in terms of labor per unit of copper output, but they concede that the "increase in scarcity of copper may be a short-term phenomenon" (Johnson, Bennett (1980: 47)) and could be alleviated by substitution.

This concession characterizes one of the disadvantages of the unit cost measure. It does not take sufficiently into account the various substitution possibilities for almost any natural resource. Furthermore, only the current scarcity, not the expected scarcity is signalled[33].

The real price of copper avoids this second drawback.

The Real Price

The real price, a scarcity measure of the resource, not of the in situ resource (Pindyck (1978b: 855)), is used by most authors.

Herfindahl (1959) deflated the copper price with the wholesale price index of the U.S. Bureau of Labor Statistics (BLS). For the period since the First World War he found no persistent price trend[34]. In an update of this study, Radetzki (1977b: 6) confirmed these results. He also could not discern a long-run price trend from 1900 to 1970, if low prices during the great depression and war time controls were neglected.

Barnett and Morse (1963: 212-213) found no trend of an aggregate price of minerals relative to a GNP deflator from the last quarter of the 19th century to the late 1950s.

Nordhaus (1974: 24) calculated the price of copper relative to the hourly wage rate in manufacturing. The relative price, defined in this way, declined considerably from 1900 to 1970. If the user cost of capital is taken as numeraire, then the real copper price fell more slowly (Brown, Field (1978: 226; 1979a: 233)).

Thus, in summary, no study shows an increase of copper scarcity measured in terms of real prices. But it depends on the choice of the numeraire, whether the scarcity remained constant or decreased.

This ambiguity is one of the major drawbacks of the real price as a scarcity index. Furthermore, this measure does not allow properly for technical progress. The relatively stable real copper price after the first World War, for instance, was at least partly due to technological changes, which made it possible to shift from selective to nonselective mining practices.

Eventually, the real price does not indicate the physical exhaustion of copper, because the metal has very close substitutes[35].

Real Copper Prices in the Postwar Period

Despite these disadvantages, the real price is preferable to the unit cost measure, because expectations in regard to costs of exploration, discoveries, mining, smelting and refining are reflected in it.

Therefore a time trend in real copper prices since the mid 1950s was fitted. The regression results, compiled in table 5-3, supplement and update the studies by Herfindahl, Radetzki, Nordhaus, Brown and Field[36].

Table 5-3:

Linear and log-linear time trend of real copper prices,
1954-1980

U.S. producer price relative to ...	Linear Form			Log-linear Form		
	Intercept	Slope	R^2	Intercept	Slope	R^2
wholesale price index	-14216 (-1.275)	7.9693 (1.4057)	.0732	-4.0393 (-0.5423)	0.0058 (0.5178)	.0845
hourly wage rate	5708.2 (2.3637)	-2.7416 (-2.2331)	.1662	21.2617 (3.0196)	-0.0079 (-2.2041)	.1627
user cost of capital	1498 (3.9198)	-0.0733 (-3.7748)	.3630	28.6840 (4.5316)	-0.0125 (-3.8996)	.3782
unit cost index	19798 (1.6428)	-9.2554 (-1.5107)	.0838	18.4766 (2.5288)	-0.0056 (-1.5213)	.0847

It is evident: from 1954 to 1980 no significant increase of the real copper price can be determined for any of the deflators (at the 5% level; the figures in parentheses denote the t-values).

If the BLS wholesale price index is used as a numeraire, then the regression coefficients of the trend variable are positive, but not significantly different from zero. This means that Herfindahl's and Radetzki's results are also valid for the last few years. Still no significant trend exists in the copper price deflated with the wholesale price index.

The choice of other numeraires indicate a declining trend, not significant for the unit cost index, but significant for the hourly wage rate in manufacturing. Thus, the declining trend ascertained by Nordhaus in 1974 continued in recent years.

Most interestingly, the highest significance level for a declining time trend is obtained for the user costs of capital in mining as a numeraire. From 1954 to 1980, the average growth rate for the capital costs in mining was 6%, but only 5% for the unit production costs and just 4.8% for the U.S. producer price.

This declining trend of copper prices in terms of the user costs of capital is the major reason for the current investment crisis in the U.S. copper industry.

In regard to any econometric model, the exercise above shows the importance of a diligent choice of the price deflator. The application of a simple producer price index or even of a more sophisticated unit cost index in explaining capacity levels, for example, is usually inappropriate. Therefore, it has been abandoned in this book's econometric model.

5.3 Implications

Empirical evidence substantiates the presumption that copper reserves and resources will last for a long time. They will continue to satisfy the long-term demand for the metal[37].

Moreover, in the near future, current reserves will most likely be supplemented by copper from polymetallic ocean bed nodule resources[38]. With some technical progress, copper deposits in the Arctic and Antarctic could be exploited[39].

If a technically useable, environmentally clean and virtually unexhaustible energy source, like controlled fusion[40] could be found, then vast stores of minerals with a very low copper yield could be exploited. Furthermore, then, not "earthbound" (Park, 1975) any more, we could leave the "spaceship earth" (Boulding, 1966) and mine copper in outer space:

- on the moon mineral deposits have been discovered according to Zuckerman (1979), and also
- within the asteroid belt the existence of metallic objects has already been proven (Gaffey and McCord, 1977).

Although some scientists are skeptical as to the future availability of mineral re-sources[41], most members of the economics profession share a moderate optimism[42].

Because the copper market appears to be far from exhaustion, the actual behav-ior of the market participants is dominated by short-term perceptions. The planning horizon of the most important copper exporters certainly does not exceed 25 years[43], a time span so small, that in praxi the problem of depletion is neglected[44].

This is one of the reasons, why Hotelling-type exhaustible resource theory[45] cannot be applied directly to describe the short-term behavior of the economic agents in the copper market.

Besides, some of the restrictive assumptions of the traditional model are not fulfilled. For example, it was shown previously that discoveries of copper deposits, yielding upward revisions of the estimated reserves, are characteristic for the world copper industry. But this increase in estimated reserves changes the basis for the calculation of Hotelling rents[46]. Also, in the copper industry, extraction costs are significant. Uncertainty pervades the industry. Government interventions certainly are not lacking[47].

Furthermore, the costs of capital and debt servicing are very high compared with the value of output in the copper industry. This implies, that the producers cannot slow down or even stop extraction, because they feel that the growth in the rental rate of the mineral exceeds the rate of interest of a comparable asset[48].

Not surprisingly, there are only a few tests of the empirical relevance of Ho-telling's model for copper and the results have not been very successful.

In an analysis of the relationship between interest rates and metal price move-ments Heal and Barrow (1980: 167) obtain the worst results for copper. Despite an extensive specification search, Smith (1981: 111) finds no acceptable model incorpo-rating a relation between copper price movements and the return on capital assets for data from 1900 to 1970. Heal and Barrow (1980) show that changes in the returns of capital assets, not their levels, appear to be the determinants of price move-ments.

This fact, however, can be explained easily with the econometric model pre-sented below. Interest rates influence industrial production, which is an important determinant of the total use of copper. And (as opportunity costs of holding copper inventories), they also influence the storage decisions of producers, consumers and merchants.

Summing Up

Although copper is an exhaustible resource, the scarcity of the metal did not increase in the last few decades. Actually, some indexes even show a significant decline.

Copper reserves and resources grew almost exponentially since the second World War, because technical progress improved exploration activities and increasingly rendered possible the mining of very low grade ore.

For most copper producers, the ratio of the reserve base to current output is far greater than their planning horizon. This is an excellent reason, why Hotelling-type exhaustible resource theory cannot be applied to explain the short-term behavior of the copper market participants.

5.4 Notes

1. Prokop (1975: 10) quotes various sources estimating this figure. Other sources claim that the copper abundance averages 55 ppm (Bowen and Gunatilaka (1977: 3), 50 ppm (Bender (1976: 11), Brobst (1979: 123) or 33 ppm (Schröder (1966: 12)).

2. Seawater contains three tons of copper in one cubic kilometer (Wenck (1969: 167)). Copper's abundance in the ocean crust is somewhat higher than in the continental crust according to sources quoted by Vokes (1976: 76).

3. For a general survey on origins, evolution and present characteristics of copper ore deposits see Tatsch (1975). This volume contains an extensive bibliography, but without economic references. For short descriptions of the geological characteristics of copper deposits see e.g. Kraume (1966), Prokop (1975: 10-11) or Gluschke et al. (1979: 45-46).

4. Copper deposits in the United States are the recorded best in the world. The U.S. Bureau of Mines conducted a series of mineral availability studies, covering also copper. For a general description of this program see e.g. Bennett, Thompson, Kingston (1977). Rosenkranz, Davidoff and Lemons (1979) appraised copper resources only. Ridge (1976) compiled a comprehensive bibliography of mineral deposits in Africa, Asia (excluding USSR) and Australasia. It includes many references to copper. Sources concerning copper deposits in the Soviet Union can be found in the bibliographies by Sutulov (1967) and Alexandrov (1980). For China see Wang (1977: 139-144). World mineral supplies in general are assessed e.g. in Govett and Govett (1976, 1977) and in the sources quoted by Brobst (1979: 117).

5. See Howe (1979: 219). Gluschke et al. (1979: 54, 62) refer to recently published literature, which indicates that the mining of lower grade ore does not necessarily lead to an increase in a deposit's copper content.

6. Cf. Whitney (1854: 234). The ore grade declined to 6.6% in 1853.

7. Prokop (1975: 52-52) reflects about the future ore grade of copper deposits, depending on the technological developments and substitution possibilities. Rampacek (1977) describes the impact of research and development in the utilization of low grade copper deposits.

8. The ore grades are compiled from various issues of the <u>Minerals Yearbook</u>.

9. For an economic explanation of this phenomenon see e.g. Herfindahl (1967).

10. Excellent guides in interpreting data on copper reserves and resources are Gluschke et al. (1979: 45-62) and United States Geological Survey (1980).

11. See <u>Mineral Commodity Summaries</u> (1981: 184-197).

12. Loc. cit., p. 184.

13. Cf. Schroeder and Jolly (1981: 6) and <u>Mineral Commodity Summaries</u> (1984: 41).

14. I assume that Tilton's (1977a: 10) estimate of 100 million tons of copper reserves in 1950 is correct.

15. The percentages in parantheses refer to 1983.

16. Calculated from <u>Mineral Facts and Problems</u> (1970: 541) and from <u>Mineral Commodity Summaries</u> (1981: 41; 1984: 41).

17. For evidence supporting this view see e.g. Bergsten (1978: 145-148), Crowson (1979) or Radetzki (1980, 1981).

18. In 1983, the total land-based resources totalled up to 1.6 billion tons of copper. Seabed nodules are estimated to contain 0.7 billion tons of copper according to <u>Mineral Commodity Summaries</u> (1984: 41).

19. See Sutulov (1979) for an evaluation of Chile's copper resources.

20. I.e. <u>Resources for Freedom</u> (1952: 36).

21. The U.S. Bureau of Mines figures are compiled from various issues of <u>Mineral Facts and Problems</u> and from the <u>Commodity Data Summaries</u>.

22. Quoted from Gluschke et al. (1979: 57 and 61).

23. In an excellent study, Streissler (1980) discriminates 12 different notions of scarcity. See also, among others, Brown and Field (1978, 1979a), Fisher (1979), Pethig (1979), Siebert (1979) or Schneider (1980) for an analysis and critique of fundamental concepts of resource availability.

24. Tilton (1977b) discusses the relation between geologic exhaustion, which is final, and economic exhaustion, which may not be final, and their relationship to estimates of resource availability.

25. Inclusion of deep sea nodules leads to a resource / consumption ratio of 243 in 1980.

26. Cf. Meadows et al. (1972). For a more general critique of this study see e.g. Beckerman (1972) or Nordhaus (1973).

27. See Meadows et al. (1972: 60).

28. Cf. also Gluschke et al. (1979: 60).

29. Siebert (1979: 410) agrees with this conclusion. If a very naive Cornucopian fits an exponential trend to the data of table 5-2, he will be pleased about a high coefficient of determination and he will readily calulate that the total quantity of copper in the earth's crust will be economically exploitable in some 2,000 years! Obviously, the high average annual growth rate of copper reserves (5.5% from 1950 to 1980) cannot be sustained indefinitely.

30. For a differing opinion see e.g. Barnett and Morse (1963: 225-227).

31. See Siebert (1981: 185-186) for a clear exposition.

32. Cf. Barnett and Morse (1963: 167).

33. For a thorough critique of the unit cost of extraction index see e.g. Brown, Field (1978: 230-232; 1979a: 220-224), Fisher (1979: 256-258) or Siebert (1979: 412). The critics of the concept are criticized by Johnson, Bell and Bennett (1980: 265-269).

34. Herfindahl (1959: 239-240) summarizes his results in more detail.

35. For more caveats see Brown, Field (1979a: 224-227) or Siebert (1979: 412-414). Additionally, in the studies quoted in the main text, I often miss an evaluation of the institutional and technological changes in the copper market, and I always miss statistical tests for the stability of the regression coefficients.

36. Sources are the U.S. Statistical Abstract for the BLS wholesale price index, and the Economic Report of the President (1982: 276) for the hourly wage rate in manufacturing. The construction of the unit cost index for mining and of the user cost of capital index is described in Appendix I.

37. This general availability of copper does of course not mean that supply security exists for all consuming countries.

38. Lenoble (1981) discusses and updates figures about deep sea nodules in the North Pacific.

39. Near Bornite, Alaska, on the Kobuk River above the Arctic Circle, prospectors recently have discovered a stupendous deposit of copper. Iskander (1973) deals with the economic impact of copper in the Canadian Arctic.

40. An official of the Princeton Plasma Physics Laboratory, where major experiments to develop nuclear fusion are conducted, told me that a fusion reactor could not be expected before the early 21st century. He guessed that even then the price of fusion energy would be 20% to 30% higher than the price of fission energy.

41. Examples are Park (1975) or Ehrlich et al. (1977).

42. To name just a few which have not been mentioned above: Netschert, Landsberg (1978: 51), Herfindahl (1967: 87-89), Banks (1976: 248), Radetzki (1977b) or Sies (1981a: 157).

43. This is e.g. CODELCO's upper limit of long-term planning (CODELCO (1982: 21)).

44. Cf. the last column in table 5-1.

45. The amount of literature on exhaustible resources grew immensely since Hotelling's classic study. A standard reference is Dasgupta, Heal (1979). Siebert (1983) presents an excellent analysis of natural resource economics in general.

46. See Arrow and Chang (1980) for an analytic treatment of this case.

47. In the last few years many extensions of the simple model have been published, which remove some of these simplifying assumptions. Devarajan and Fisher (1981) survey these developments.

48. For this and additional points see Mikesell (1979: 310-313).

PART II

Chapter 6: Copper Market Models

This chapter briefly surveys existing econometric and other quantitative copper market models. The final section describes the main features of a new econometric copper market model, which the rest of the book will discuss in detail.

6.1 Econometric Models

Like any model, an econometric model of the copper industry is a tool and thus, is useful only with regard to a certain range of applications.

Econometric copper market models have been tailored for a variety of different yet interrelated purposes, mainly for forecasting and for simulation experiments[1].

Forecasting

Many of the econometric models to be mentioned in this chapter have been used for short and long term forecasting exercises. Forecasting prices and export earnings of producing countries in the short run, projections of supply, demand, and prices in the long run, are among the major applications.

Forecasts are often produced commercially. Firms that generate and sell such forecasts include Wharton Econometric Forecasting Associates (WEFA), Charles River Associates[2], Chase Econometrics[3], the Commodities Research Unit[4] and the Rio Tinto Zinc Limited[5].

WEFA, previously owned by the University of Pennsylvania, has the best track record in forecasting, based on a long experience in the econometric modeling of mineral markets. Apart from the studies mentioned later in this section, several in-house models of the world copper industry have been developed by WEFA in recent years, the latest being Kellman (1978) and Pobukadee (1980)[6].

Besides the commerical firms mentioned, the Economic Analysis and Projections Department of the World Bank and the research staff of the International Monetary Fund have developed econometric commodity market models[7], among them copper studies by Thiebach (1976), Tims, Singh (1977), Richard (1978), Thiebach, Helterline (1978) and Hwa (1979).

Based on a study by Khanna (1972), Thiebach (1976), Tims, Singh (1977) and Thiebach, Helterline (1978) developed simple two equation models to explain the LME price and the total world consumption. These models were designed to aid copper price projections and to supplement expert opinion. Soon, these forecasting devices were superseded and supplemented by more complex models.

Richard (1978) developed a dynamic continuous time model of the world copper industry, based on a stock-flow hypothesis of price formation, which attempted to represent a non-tâtonnement process. Hwa (1979) analyzed various approaches to price determination in international commodity markets, including the copper market, and formulated a dynamic disequilibrium model of price adjustment[8].

Currently, Castelli is working on a new large-scale econometric copper market model for the World Bank[9].

Simulation Experiments

While the models mentioned above have predominantly served to perform copper market forecasts, other econometric models are mainly designed for policy simulation exercises. The line between forecasting and simulation models is of course very blurred and to a considerable degree arbitrary. Discussing a huge number of the models systematically, it may be useful as a didactical device.

Dynamic simulations with econometric copper market models have been performed to serve the following purposes:
- to assess policy proposals for copper producers, e.g. to evaluate the effects of potential supply restrictions,
- to estimate the impact of copper production from deep sea nodules,
- to determine the consequences of environmental and other regulation on the copper industry,
- to appraise the substitution and recycling of the metal,
- to analyze the impact of copper on local economies, and on the world economy, especially
- to assess the impact of copper price stabilization schemes.

The rest of this section surveys briefly the models that have been applied to the above policy simulations.

Assessment of Producer Policies

The assessment of policy proposals for producer countries, especially for the developing economies, has been one of the main purposes of econometric commodity market modeling.

As a salient example, Fisher, Cootner and Baily (1972) analyze the impact of differential copper production policies of the Chilean government on the world market. This classic econometric model of the world copper market has had an impact on future copper market modeling, which can not be overestimated[10]. It is based on many forerunners, among them the first reasonable medium scale model of the world copper market, developed by Behrman (1970)[11].

One of the possible producer policies mentioned in Chapter 4 is cartelization. Carlson (1975), Charles River Associates (1976) Underwood (1976, 1977), Pindyck (1978a) and Wagenhals (1984d, e) analyze possible gains to developing copper exporters in case of a cartelization and conclude that the chances for a cartel appear slim.

Ocean Floor Nodules

The potential production of copper from polymetallic ocean floor nodules is likely to have impacts on copper prices and on export earnings of land-based producers of the metal. Adams (1973a, 1975, 1978c, 1980) published a series of ground breaking papers dealing with the effects of nodule exploitation on the world copper market. The impact on the revenues of the developing copper exporting countries was a main focus. He finds that the effects of nodule production on the copper price are probably small, but that the foreign exchange earnings of copper producing developing economies are likely to be reduced significantly due to the substitution of ocean floor production for land-based production.

Environmental Regulations

The second chapter gave some examples of how in the United States environmental regulations have affected the copper industry. It is not surprising that the impact of pollution abatement and other regulations have also been analyzed along with the help of econometric models of the U.S. copper market[12].

Substitution and Recycling

Substitution of copper and recycling of the metal are implicitly considered by many of the larger econometric copper market models. A few studies, however, deal more explicitly with these topics.

Pickard and Krumm (1977) attempt to analyze the interaction between the copper and aluminum market with an econometric model. Taylor's (1979) model of the U.S. copper industry also includes the examination of inter-metal complementarities. Brown, Field (1979b) estimate factors affecting the substitution of depletable resources, including copper. Anderson and Spiegelman (1976) assess the impact of the U.S. Federal Tax Code on the resource recovery and analyze the extent to which tax subsidies effect the recycling of a variety of materials, including copper.

Based on her 1979 dissertation, Slade (1980a, b, c) published several articles which analyze the possible impacts of policies on recycling and the effectiveness of alternatives to recycling. Studies by Bonczar, Tilton (1975), Synergy, Inc. (1975, 1977) or Staloff (1977) deal also with problems of recycling and substitution with the help of econometric models of the U.S. copper industry.

Impact on Local Economies and the World

One of the major subjects of econometric copper market studies is the impact of copper on local economies.

Lira (1974, 1980) and Lasaga (1979, 1981) analyze the Chilean experience; Obidegwu (1979) and Obidegwu, Nziramasanga (1981a, b) deal with the Zambian case.

Mahalingasivam (1969) models the Canadian refined copper market. Sawyer, Obidegwu (1979) incorporate a model of the Canadian copper industry in a macroeconometric model of Canada and examine the effects of LME price changes on the Canadian economy. This paper illustrates how international price feedbacks and transmission mechanisms can be captured by incorporating a copper market model into a national econometric model.

Ideally, total impacts of copper and other commodity price movements should be considered in the context of a global model, like, for instance, the Project LINK[13].

In a series of excellent papers, Adams (1973b, 1977, 1978a, 1979a, b) develops and pursues his idea of COMLINK. This system is derived from an expansion of the LINK system by integrating small commodity market models to evaluate the impact of commodity price movements on the world market.

Price Stabilization

Copper is one of the main candidates for stockpiling agreements. Experience about the practical application of international buffer stocks does not exist for copper. Therefore policy makers have looked for the guidance of econometric and other quantitative models to assess the potential impact of a copper price stabilization scheme.

In an econometric simulation study, Smith (1975) evaluates international copper buffer stocks. His report suggests that copper price instability could be cut in half at little or no net costs. His requirements for such results provide: that the only objective of an adequately financed buffer stock authority is the reduction of price fluctuations around a long-term equilibrium level, that export and production quotas are not included in the agreement, and that price fluctuations are allowed to such an extend that private storage and the future markets are not influenced.

In a report prepared for UNCTAD, Charles River Associates (1977) analyze the feasibility of a copper price stabilization using a buffer stock or supply restrictions with an econometric model of the world copper market. The study's main conclusion is that fixed price band rules lead only to a small decline in price instability.

Perlman (1977) compares the performance of national and international stabilization schemes for copper by performing econometric simulations with the Commodities Research Unit's quarterly model of the world copper market. His main results indicate, that national stabilization schemes are less efficient than an international scheme, but that even an international buffer stock authority would need facilities at least 2.5 times higher than previously assumed by UNCTAD.

The same model is used by Bhaskar, Gilbert and Perlman (1978) to evaluate the financial profitability of copper price stabilization. Under alternative assumptions, their stabilization schemes always yield a negative net present value. They also conclude that the size of the facility proposed by UNCTAD is insufficient to counter the high copper price volatility.

All stabilization studies mentioned above have a major drawback for an independent researcher, who would like to check the results: they do not include descriptions of the estimated equations except in a very vague, verbal form.

According to professional model sellers, this is due to the time consuming effort and the considerable expenses necessary to develop a realistic econometric model of the copper market.

On the other hand, because commodity price stabilization is not just a pure economic problem, but includes many political aspects, and in view of the differential results mentioned above, the unmasking of the underlying assumptions of these models is very desirable.

As a comfortable exception to other copper price stabilization studies mentioned, the papers of Behrman and Tinakorn[14] satisfy this criterion. In several econometric studies, these authors evaluate extensively the price stabilization aspects of UNCTAD's plan for an Integrated Commodity Program, proposed in December 1974. Despite some provisios, they approve of a price stabilization program for the ten core commodities including copper.

Summing Up

This section has showed the extent of econometric copper market modeling.

While Labys (1978), in his inventory of commodity market models, lists only five econometric copper market models, and Hu, Zandi (1980) assert that only three (!) econometric copper market models have been published, I have counted more than 80 papers based on econometric copper market models.

These models have been used for forecasting and for simulations in order to provide a prerequisite and basis for planning and to serve as an instrument to improve the understanding of the copper market mechanism in general.

Looking at econometric copper market models alone does not always make always clear, if at present, econometrics is a branch of science or more akin to alchemy (Hendry (1980)). Yet even a fervent critic of econometric copper market models like Banks, admits that "it remains true that together with a knowledge of economic theory, history, and the mechanics of the particular market, econometric techniques are invaluable for studying various aspects of these markets" (Banks (1977: 12)).

To complete the description of quantitative copper industry models, the next section briefly surveys some non-econometric approaches.

6.2 Other Copper Industry Models

Process Optimization Models

Apart from econometric approaches, process optimization models using mathematical programming techniques have been used frequently to model the copper industry[15].

These approaches usually concentrate on the U.S. copper industry and use very detailed information, including site specific data. The world market and its institutional features are, however, often neglected. Adjustment processes based on changing expectations, for instance, are generally ignored[16].

Hibbard, Soyster and Gates (1980a, b) developed a disaggregated engineering supply model of the U.S. copper industry[17]. They used this linear programming model to forecast prices and production up to the year 2000 and to evaluate whether the U.S. industry can absorb costs of environmental regulations.

While this model assumed a competitive market, Soyster and Sherali (1981) recognized the oligopolistic features of the U.S. market. An application of their revised model led to significantly different results, although they used the identical data base as the 1980 study.

Patterns in the international minerals trade, including the copper trade, have also been analyzed with a linear programming approach[18]. The studies by Tilton (1966), Kovisars (1975) and Whitney (1976) are examples.

Williams, Brodice and Poulter (1973) use linear programming to evaluate production strategies in a group of copper mines.

The copper model of Dammert (1977, 1980) and Dammert, Kendrick (1979) is solved as a multiperiod mixed integer linear program which minimizes total investment costs subject to capacity, materials balance and other constaints.

Other Approaches

The proliferation of econometric and process optimization models of the copper industry as indicated above began in the early 1970s. Quantitative assessments of prospective copper market developments were needed many years before the seventies decade.

Because of the material shortages during and after the Second World War, U.S. President Truman created the Paley Commission in 1952. Volume II of the report published by this Commission (Resources for Freedom (1952)) included copper demand projections based on aggregate projections of the U.S. economy's growth and of trends in substitution[19].

Malenbaum (1973, 1977) developed another long-term projection of materials requirements. Based on estimates of the gross domestic product and of the intensity of use as well as on the "judgment of the research worker" (Malenbaum (1972: 2) he obtained figures for the required amount of refined copper for ten subdivisions of the world in the year 2000.

The U.S. Bureau of Mines regularly publishes copper demand projections. These conditional forecasts are based on ordinary least squares estimates and on a macroeconometric model of Data Resources Inc.[20].

The Bureau uses also a resource base model, the so-called "Supply Analysis Model" (SAM), a minerals availability system methodology, which does cover the copper industry[21]. Rosenkranz, Davidoff and Lemons (1979) used this model to examine problems of resource availability in the United States.

In an engineering study with some econometric elements, Foley (1979) and Foley, Clark (1981) developed short and long term supply schedules for the primary U.S. copper industry based on site- and input-specific cost data. Silverman (1977) and Hu (1978) also pursued an engineering approach and analyzed material flows for the U.S. copper industry.

Industrial and system dynamics methods have been applied to the copper industry by Ballmer (1960), Schlager (1961) and the Pugh-Roberts Associates (1976)[22].

Finally, autoregressive moving average models have been used to forecast copper prices. Chu (1978) applies ARIMA techniques for short-run projections of monthly prices of copper and other commodities. Mingst and Stauffer (1979) explain the LME copper and tin prices with an univariate Box-Jenkins technique[23].

Adding up

The last two sections showed that considerable reseach effort has been dedicated not only to econometric copper market modeling, but also to other approaches, mainly by studies using mathematical programming techniques.

The next section briefly introduces a new econometric model of the world copper industry. Based on the tradition of previous studies, it includes several new features.

6.3 A New Econometric Model

The following chapters of this book describe in detail a new econometric model of the world copper market, which has these special features:

1. It is the most highly disaggregated econometric copper market model currently available. It therefore allows the analysis of many country specific problems.

2. Primary supply functions are not derived from the common partial adjustment approach, but from restricted profit functions, adding to the model's realism.

3. Copper mine production capacities are explained endogenously. They are derived from the hypothesis that producers act to maximize their discounted net cash flow. This feature allows, for example, assessment of the effects of alternative tax systems on investment behavior.

4. A dynamic stock disequilibrium approach is used to determine the LME spot copper price. The LME copper futures price is explained endogenously.

5. Producer price equations are introduced not only for the United States, but also for Canada and Chile. This means that alternative producer price policies cannot only be simulated for the United States, as in most copper market models, but also for Canada and Chile.

6. The modeling of private inventory behavior uses a rational expectations approach.

   In an analogy to Part I, the next three chapters model the copper supply, demand, and price formation.

6.4 Notes

1. Wagenhals (1983c, 1984a) describes limitations and possibilities of econometric non-ferrous metal market models in general. A recent bibliography (Wagenhals, 1983b) lists more than 150 of these models. For a state-of-the art survey of mineral markets modeling in the late 1960s and 1970s see e.g. Behrman (1968), Burrows (1975) and Labys (1977).

2. Copper studies based on various versions of CRA's econometric copper market model are Charles River Associates (1970, 1971, 1973, 1976, 1977, 1978), Burrows (1975), Burrows, Lonoff (1977), Lonoff (1978, 1980), Lonoff, Reddy (1979), Klass, Burrows, Beggs et al. (1980).

3. Chase Econometrics does not publish its econometric copper market model in any detail. According to R.G. Adams, vice president of Chase Econometrics, the copper market model is similar to Chase Econometric's zinc market model (see Adams, R.G. (1981)), taking into regard several special features of the copper industry. For copper market forecasts cf. Adams, R.G. (1979).

4. The econometric research of the Commodities Research Unit Ltd. (CRU), an international consultancy specializing in the minerals industry, has not been written up in any detail. (Personal communication with D. Gilhespy, Senior Consultant, Copper, CRU, London.) Allingham (1977) and Allingham, Gilbert (1977) provide a very scant description of CRU's long-term copper model.

5. The Rio Tinto Zinc Limited does econometric work of the base metals, especially of copper. The model is based on the Fisher, Cootner, Baily (1972) approach and is highly disaggregated on the supply side. (See Tinsley (1977: 705) and Crowson

(1979: 167-169).) The model and its results are confidential and are used for internal purposes of the Rio Tinto Zinc only. (Personal communication with P.C.F. Crowson of the Rio Tinto Zinc Corporation Limited, London.)

6. Ogawa's study (1982a, b) is based on Pobukadee's approach, but he assumes different formats for the supply and demand functions and includes two investment functions for the United States and the rest of the market economies.

7. See e.g. World Bank (1981).

8. W. Labys currently develops also a disequilibrium model of the world copper market. For preliminary results see Labys (1980b) and Labys, Kaboudan (1980). Another disequilibrium approach, restricted to the estimation of the U.S. demand for refined copper, is applied by MacKinnon and Olewiler (1980).

9. See Castelli (1982) for first results.

10. Virtually all later econometric copper market models rely in one way or another on this model. Almost identical, only slightly modified or aggregated versions are the models by Underwood (1976, 1977), Pindyck (1978a), Carlson (1975), Synergy, Inc. (1975, 1977), Hansen (1978), Lee (1980) and Lee, Blandford (1980a, b).

11. Other important precursors of the Fisher, Cootner, Baily model include Newhouse and Sloane (1966), who modelled supply functions for the most important copper producers, and Ertek (1967), who dealt with the world demand for copper. For earlier demand studies see Bradley (1948) and Thomas (1962). Banks (1969) also worked on an econometric model of the world copper market.

12. See Arther D. Little, Inc. (1976; 1978a), Charles River Associates (1971), Stevens (1977), Hartman, Bozdogan, Nadkarni (1979) and the Massachusetts Institute of Technology (1979). Hartman (1977 b), Mikesell (1979) and Hu, Zandi (1980) compare the models of Arthur D. Little, Inc., Charles River Associates and Fisher, Cootner, Baily. An important drawback of many of these approaches is the assumption that the LME price is given exogenously. From the above description of copper pricing it should be clear, that this hypothesis is not very reasonable, even if only the U.S. market is considered: the U.S. producer price and the LME price are linked by arbitrage transactions at COMEX. The trade flows between the United States and the rest of the market economies are correlated with the London price.

13. For a description of Project LINK see e.g. Waelbroek (1976) or Sawyer (1979).

14. Cf. Behrman, Tinakorn (1977, 1978, 1979, 1980), Behrman (1978) and Tinakorn (1978). See also the papers by Ghosh, Gilbert and Hughes Hallett (1981, 1982).

15. Sparrow and Soyster (1980) survey some process optimization models used in the minerals industries.

16. Adams and Behrman (1981: 391-392) discuss advantages and disadvantages of programming models.

17. Production outside the United States is included in considerably less detail. A demand submodel is based on Arthur D. Little (1978).

18. International trade flow modeling is surveyed by Demler and Tilton (1980).

19. Cooper (1975) and Baak (1978) look at the report of the Paley Commission again, more than two decades after its publication.

20. Cammarota, Mo, Klein (1980) and Mo, Klein (1980) describe the methodology used for the Bureau's projections in more detail. See Sousa (1981: 78-80) for a comparison of several long-term copper consumption forecasts based on an intensity-of-use approach.

21. Davidoff (1980) evaluates the SAM system and its analytical capabilities.

22. Strongman, Killingsworth and Cummings (1977) summarize the latter model.

23. Other projections of copper prices and quantities not based on econometric or mathematical programming methods are included, for instance, in Ridker (1972), Takeuchi (1975), Banks (1976), Rohatgi, Weiss (1977) and Radetzki (1983). Varon and Gluschke (1977) survey mineral commodity projections as a tool for planning.

Chapter 7: Mine Production and Capacities

7.1 Overview

The last chapter gave a picture of the vast amount of research effort dedicated to the modeling of the world copper market.

The following chapters contain a write-up of a new econometric model which is based on many precursors described earlier, but which has also some distinguishable new features. This chapter deals with the specification and estimation of equations explaining the primary supply and copper mine capacities of the eight most important copper producing countries and two supranational regions.

Unlike the traditional models where a simple partial adjustment hypothesis is used to justify a Koyck-type supply function, this study starts from the explicit economic hypothesis that copper producers maximize their profits given their capacity restrictions. Via factor demand functions and restricted profit functions the application of Hotelling's lemma immediately leads to primary supply functions which are estimated for eight countries and two supranational regions.

The copper mining capacities are neglegted in virtually all econometric copper market studies. This is due to the deficient capacity data published in trade jour-nals. However, thanks to a major American copper producer, I was able to use unpub-lished confidential data collected by its statistical department, which turned out to be far more reliable.

In this study, copper mining capacities are derived from the hypothesis that copper producers maximize their discounted net cash flow. Capacities depend on prices, output, exchange rate developments, user costs of capital and on risk con-siderations. This neoclassical approach works quite well. Structural breaks, how-ever, in the investment behavior of some developing producers due to nationaliza-tions or due to changes in the majority ownership of the mines necessitate the in-troduction of dummy variables.

The equations for the primary supply of the most important copper producers are derived from economic theory in the following section. For typical equations, the estimation results are then explained and interpreted. The complete results are compiled in Appendix III.

7.2 Primary Supply: Theory

This section deals with the derivation of behavioral equations explaining the primary copper supply, i.e. the supply of copper from newly mined ores.

Almost every econometric model faces a tradeoff between model complexity and data. This is also true for this world-wide model of the copper industry, where data for the developing economies in general and figures for inventories and capacities in the industrialized countries are notoriously deficent, when available at all. Detailed time series of capital, labor, energy and material input data were not available for most of the copper producing countries. Due to these data limitations, I reverted to a simple theoretical approach and assumed that a generalized restricted Cobb-Douglas production function sufficiently describes the process of copper mining:

$$q(V) = AV^{\alpha}K^{\beta},$$

where q is the mine production of a country,

            V is a bundle of variable inputs,

            K denotes the mine production capacity,

            A, α and β are constants.

In accordance with the considerations in chapter 2 the capacity is assumed to be given in the short run.

In a competitive setting, the first order condition for profit maximization is

$$\alpha pAV^{\alpha-1}K^{\beta} - c_V = 0,$$

or, equivalently,

$$\alpha pq = c_V V,$$

where c_V is the unit cost of the bundle of variable factors and p is the output price.

Thus,

$$V^{\alpha-1} = c_V (A\alpha p)^{-1} K^{-\beta},$$

and solving for V yields the factor demand function

$$V = \left(\frac{p}{c_V}\right)^{1/(1-\alpha)} (\alpha A)^{1/(1-\alpha)} K^{\beta/(1-\alpha)}.$$

The restricted profit function therefore is

$$\pi(p, c_v; K)$$

$$= A^{1/(1-\alpha)} (\alpha^{\alpha/(1-\alpha)} - \alpha^{1/(1-\alpha)}) p^{1/(1-\alpha)} c_v^{\alpha/(\alpha-1)} K^{\beta/(1-\alpha)} - c_k K,$$

where c_k is the unit cost of capital.

According to Hotelling's lemma (the "derivative property"), the producer's supply function is the partial derivative of the restricted profit function in respect to the output price:

$$q(p, c_v; K) = \frac{\partial \pi(p, c_v; K)}{\partial p},$$

when output and input prices are positive (see e.g. Varian (1978: 31)). With this lemma, the supply function follows immediately:

$$q(p, c_v; K)$$

$$= \frac{1}{1-\alpha} A^{1/(1-\alpha)} (\alpha^{\alpha/(1-\alpha)} - \alpha^{1/(1-\alpha)}) (\frac{p}{c_v})^{\alpha/(1-\alpha)} K^{\beta/(1-\alpha)},$$

or

$$\log q = \alpha_0 + \alpha_1 \log(\frac{p}{c_v}) + \alpha_2 \log K,$$

where the definitions of α_0, α_1 and α_2 are obvious.

This equation has been estimated for the eight most important copper producing countries and for two supranational regions. Section 7.3 deals with the results.

Comparison with the Traditional Approach

Compared with the simple partial adjustment assumption applied in virtually all econometric copper market models from Newhouse, Sloan (1966) to Fisher, Cootner, Baily (1972) to Pobukadee (1980), the current approach is superior not only in terms of higher coefficients of determination and other statistics, but also

- because the partial adjustment hypothesis in regard to copper mine production often leads to estimates of the long-run price elasticities, which appear unacceptably high[1], and

- because the estimated equation is based on ad hoc dynamics only instead of being based on an optimization approach using microeconomic theory. Therefore, it can be difficult to interpret the parameter estimates unambiguously, because it is not clear what is held constant during the transition from short to long run[2].

Structural Change

Econometric copper market models are sometimes criticized because of the stipu-
lation that copper producing countries act as if they maximize profits.

Indeed, state copper mining enterprises in developing countries have more
complex goals than only pure profit maximization. The Zambian government, for
example, asserts to maximize output and profits (Faber, Potter (1971: 110)). Other
possible targets of copper producers which have been mentioned in the literature
include maximization of foreign exchange earnings, of employment, of taxes and
royalties from mining or increasing the world market share.

One might, however, argue that these critics overlook the fact that the assumption of profit maximization is an approximation to reality which is compatible with
rather different optimization approaches (Leland (1980)), and that countries might
be forced to follow a profit maximization target if they want to attract foreign investors and obtain funds from commercial banks.

Anyway it has to be tested statistically whether a nationalization or a new 51%
ownership in equities of copper mining companies led to a significant change in the
response of the copper supply in the developing countries. This task has been
neglected in econometric models of the copper industry up to now. I therefore tested
for a structural break in the year of nationalization or the beginning of majority
ownership in Chile, Peru, Zaire and Zambia with an F-Test ("Chow-Test").

The results showed that for the primary supply functions no structural break
can be ascertained between the periods before and after nationalization (or majority
ownership). Therefore I felt that the assumption of short-run profit maximization
can still be justified as hypothesis in the explanation of primary copper supply,
and accordingly I estimated the supply function derived above.

7.3 Primary Supply: Estimation

The coefficients of the primary supply equations derived in the last section
were estimated for the eight most important copper mining market economies (Canada,
Chile, Peru, Philippines, South Africa and Namibia, United States, Zaire and
Zambia), and for two regions consisting of the rest of the Western World ("other
market economies") and the centrally planned countries.

Market Economies

As typical example for a primary copper supply equation of a developing country, table 7-1 presents the estimated equation for Zambia's mine production.

Table 7-1

Estimated mine production equation, Zambia

$$\log(QMZM) = -.234 + .908 \log(QMZMC) + .103 \log((PCULME*REXZM)/(REXUK*COSTZM)),$$
$$(-.502)(15.3)(3.44)$$

$$\bar{R}^2 = .904, \qquad\qquad DW = 1.80,$$

where

QMZM   = Copper mine production, Zambia, in 1000 tons,

QMZMC = Copper mine production capacity, Zambia, in 1000 tons,

PCULME = Average annual price of copper, electrolytic copper, wirebars, London Metal Exchange, cash, Pound sterling per ton,

REXZM = Exchange rate, Zambian Kwacha per U.S. Dollar,

REXUK = Exchange rate, Pound sterling per U.S. Dollar,

COSTZM = Unit cost of mining index, Zambia, 1975 = 1.00.

The figures in parantheses are the t-statistics, $\bar{R}^2$ denotes the adjusted coefficient of determination and DW means Durbin-Watson coefficient. The signs of the estimated coefficients are as expected from economic theory; all coefficients are significantly different from zero and have magnitudes which are a priori plausible.

The complete estimation results are compiled in Appendix III. Appendix II lists the variables along with short descriptions and their sources. General remarks on methodology and data can be found in Appendix I.

Appendix III shows, that not only the coefficients of Zambia's mine production equation, but the coefficients of all estimated primary supply equations have the expected signs and, with the exception of two price elasticities of supply, all coefficients are significantly different from zero (at the 5% level).

The small copper producers' output is highly autocorrelated and depends on prices only to a limited extent. South Africa's price elasticity is not significantly different from zero, because, as section 4.1 showed, she consumes the main part of her copper production domestically and her links to the world market are comparably weak.

Serious labor unrests in Canada and in the United States as well as the Katanga war and its aftermath in Zaire have to be accounted for by dummy variables. These, however, are the only dummy variables in the primary supply equations, as compared to most econometric copper market models with a similar degree of disaggregation.

To complete the discussion on estimated primary supply equations of the market economies, table 7-2 shows the equation explaining the U.S. primary copper output as an example for an industrialized copper producing country.

Table 7-2:

Estimated mine production equation, United States

$$\log(QMUS) = -.693 + .873 \log(QMUSC)$$
$$(-1.38) \quad (22.4)$$

$$+ .202 \log(PCUUS/COSTUS) - .438 \ DSTUS,$$
$$(3.90) \qquad\qquad (-14.2)$$

$$\bar{R}^2 = .96, \qquad\qquad DW = 1.90,$$

where

QMUS   = Copper mine production, United States, in 1000 tons,

QMUSC = Copper mine production capacity, United States, in 1000 tons,

PCUUS = Average annual price of copper, electrolytic copper, wirebars, U.S. producers (f.o.b. refinery), U.S. $ per ton,

COSTUS = Unit cost of mining index, United States, 1975 = 1.00,

DSTUS = Dummy variable for major strikes in the U.S. copper industry.

Centrally Planned Economies

For the centrally planned economies no time series data on copper mine produc- tion capacities are available.

In contrast to the market economies, the mine output in the centrally planned economies has been expanding rather steadily until the late 1970s. The continuous growth of copper mine production in these countries suggests the assumption that their output follows a simple time trend and it depends on the general level of industrial activity. Of course, this is a specification error due to missing data.

The Cochrane-Orcutt method was selected to account for a first order autoregressive error process due to this mis-specification. The estimate of the autocorrelation coefficient amounted to .63, with a t-statistic of 4.1.

Finally, the primary supply equation for the centrally planned economies estimated with the Cochrane-Orcutt procedure is presented in

Table 7-3:

Estimated mine production equation, centrally planned economies

$$\log(QMEW) = \underset{(22.2)}{4.48} + \underset{(2.78)}{.681 \log(T)} + \underset{(.817)}{.167 \log(IIPEW)},$$

$$\bar{R}^2 = .967, \qquad DW = 1.98,$$

where

QMEW = Copper mine production, centrally planned economies, in 1000 tons,

IIPEW = Index of industrial production, centrally planned economies, 1975 = 100,

T = Time trend, 1963 = 10, increasing (decreasing) by 1 each succeeding (preceding) year.

7.4 Mine production capacity: Theory

The market countries' total copper mining capacity is disaggregated in the eight most important producers' and the rest of the market economies' capacity, according to the partition of primary supply. The most important factors influencing the level of copper production capacity include current and lagged copper prices and the user cost of capital, which depends - inter alia - on the purchase price of capital, the interest rate, the rate of depreciation, the corporate tax rate for mining enterprises and the depreciation method.

Furthermore, a risk variable, defined as a five year sample variance of annual LME copper prices converted to U.S. dollars and adjusted for general price level changes, has been introduced.

Based on the information compiled in chapter 5, the reader possibly expects to find a recognition of the fact that copper is an exhaustible resource in the equations explaining the copper mine production capacity. Indeed, I experimented with various formats which recognized the existence of a resource constraint.

Declining ore grades, for instance, which I collected for many copper mines, influenced the copper mining capacity only insignificantly, although generally the sign of the estimated parameter was correct.

The estimated coefficients in specifications of the capacity equations which considered the existence of a resource constraint using dynamic programming methods, proved to be very sensitive to small changes in the objective function or in the ex- ogenous variables and led to unrealistic results.

The discussion in chapter 5 suggested several explanations for this phenomenon that the Hotelling-type exhaustible resource theory cannot be applied to the short- term behavior of the copper industry.

Therefore, in the conclusive version of the model, I omitted these approaches, because I believe that the resource constraint for copper does not play a very im- portant role in the producers' decisions. The present world copper reserves are very high and they are even likely to increase considerably in the near future because of deep sea nodule mining, such that copper producers do not feel a threat of exhaustion. Indeed, in interviews with copper industry officials, I was assured that the resource constraint does not influence decisions about investments in the copper industry significantly. According to their view, cost considerations, especially concerning unit costs of capital, play a far more important role.

Next, I shall formally derive the capacity equations which have to be estimated.

Suppose that a competitive copper producer maximizes his discounted net cash flow. Then his marginal discounted net revenue is

$$\frac{1}{r}((1-t)Q_k p - c_k d) + t \sum_{i=1}^{\infty} (1+r)^{-i} D_i ,$$

where t is the corporate tax rate,

Q_k is the marginal product of capital,

p is the copper price,

c_k is the purchase price of an additional unit of capital services,

d is the rate of economic depreciation of capital goods per period,

D_i is the increase in depreciation charges for tax purposes in period i, and

r is the interest rate[3].

Let d_i be the tax depreciation permitted on an investment of one dollar i peri- ods after the investment has been made, and

$$B := \sum_{i=1}^{\infty} (1+r)^{-i} d_i$$

the discounted value of depreciation charges resulting from a current dollar of capital expenditures.

In most copper producing countries straight line depreciation is ordained for copper companies, but sometimes the sum of the years' digits depreciation method can also be used.

In the case of straight line depreciation[4]

$$B = \frac{1}{r\tau}(1 - \exp(-r\tau)),$$

and in the case of sum of the years' digits depreciation

$$B = \frac{2}{r\tau}(1 - \frac{1}{r\tau}(1 - \exp(-r\tau))),$$

where τ is the lifetime for tax purposes. In any case, the discounted net revenue is

$$\frac{1}{r}((1-t)Q_k p - c_k d) + tc_k B(1+\frac{d}{r}).$$

Capacity will be increased, if the discounted revenue stream from an additional unit of capacity is greater than the purchase price, i.e. if

$$\frac{1}{r}((1-t)Q_k p - c_k d) + tc_k B(1+\frac{d}{r}) > c_k,$$

or, equivalently, if

$$Q_k p > c,$$

where $c := c_k(r+d)(1-tB)(1-t)^{-1}$ is the user cost of capital.

The user cost of capital is the implicit rental price for a unit of capital and depends on the price of capital, the interest rate, the depreciation charge, the corporate tax rate and the depreciation method used.

For the copper producers I generally assumed straight line depreciation, which is compulsory in most of the countries. In South Africa and Namibia the diminishing balance method could also be used. For the United States, the sum of the years' digits depreciation method has been assumed. In the United States, investment tax credits and the Long Amendment, which stated that the tax credit had to be deducted from the depreciable base in 1962 and 1963, had to be incorporated.

Thus, the U.S. equation for the user cost of capital is

$$c = c_k(r+d)(1-s-t(1-ms)B)(1-t)^{-1},$$

where s is the investment tax credit rate,

m is a dummy variable for the Long Amendment which is 1 in 1962 and 1963 and 0 otherwise, and

B is calculated according to the sum of the years' digits depreciation method.

Now, suppose that a copper producer determines the desired capacity to maximize discounted profits subject to a Cobb-Douglas production function. This implies

$$K_t^* = B \frac{P_t}{c_t} q_t ,$$

where K_t^* is the desired level of capacity,

B is the constant elasticity of mine production with respect to capacity,

P_t is the copper price,

c_t is the user cost of capital, and

q_t is the copper mine production.

I assume that the capacity is proportional to the capital stock and that the capital stock is proportional to the flow of capital services.

Because of the lags in the completion of capacity changes, the actual level of capacity usually differs from the desired level of capacity. Assuming a partial adjustment of the actual to the desired level of capacity[5],

$$K_t - K_{t-1} = \lambda (K_t^* - K_{t-1}) ,$$

it follows,

$$K_t = \lambda B \frac{P_t}{c_t} q_t + (1-\lambda) K_{t-1} ,$$

or

$$K_t = B_o + B_1 K_{t-1} + B_2 \frac{P_t}{c_t} q_t ,$$

where $B_o = 0$, $B_1 = 1 - \lambda$, and $B_2 = \lambda B$.

Adding a five-year sample variance of deflated prices, R, as a proxy for the price risk term, gives eventually

$$K_t = B_o + B_1 K_{t-1} + B_2 \frac{P_t}{c_t} q_t + B_3 R_t .$$

The next section deals with the estimation results for this equation.

7.5 Mine production capacity: Estimation

Results

The parameters of the equation derived above were estimated for the most important Western copper producers (Canada, Chile, Peru, Philippines, South Africa and
Namibia, United States, Zaire and Zambia) and for the rest of the market economies
together.

Again, the estimated equation for Zambia is presented as an example for the
mine production capacity equations of a developing copper exporting country:

Table 7-4:

Estimated mine production capacity equation, Zambia

$$QMZMC = 80.7 + .896 \ QMZMC_{-1}$$
$$(2.57) \quad (23.1)$$

$$+ .000705 \ (PCULME*QMZM*REXZM/(REXUK*UCZM))$$
$$(.694)$$

$$- .258 \ R,$$
$$(-1.02)$$

$$\bar{R}^2 = .971, \qquad\qquad DW = 1.80,$$

where

R   = Five year sample variance of PCULME/(REXUK*PCIF),

PCIF = International price index, unit values of manufactures (SITC 5-8), 1975 =
        100,

UCZM = User cost of capital index, mining, Zambia, 1975 = 100, and

QMZM, QMZMC, PCULME, REXZM and REXUK are explained in table 7-1.

Table 7-4 shows that Zambia's lagged capacity level is the main determinant of
the current level. The mine capacity adjusts only very slowly to price changes. Zambia appears to be risk-averse, but the coefficient of the "risk variable" R is not
significantly different from zero.

These results are more or less typical for the mine production capacities of
the developing copper producing economies.

Although developing countries always adjust to price changes in the direction expected a priori, mine capacities generally adapt very slowly and only weakly. In contrast to these results, the capacity changes of the main copper producers among the industrialized market countries, namely of Canada and the United States, depend very highly on price changes: the price elasticities are significantly different from zero at the 1% level. Thus, the industrialized copper producers react far more on price incentives than the developing economies.

Confirming this assertion table 7-5 presents the example of the U.S. copper mine capacity equation. Dummy variables take into regard that copper mine expansions did not take place during the 1967 strike, and that investments were postponed from the recession year 1975 to the year after.

Table 7-5:

Estimated mine production capacity equation, United States

$$QMUSC = 31.6 + .921 QMUSC_{-1}$$
$$(.843) \quad (43.1)$$

$$+ .00527 (PCUUS*QMUS/UCUS)$$
$$(3.76)$$

$$- .0055 R + 123 D67 + 109 D75/76,$$
$$(-.0173) \quad (4.02) \quad (5.28)$$

$$\bar{R}^2 = .990, \quad DW = 2.36,$$

where

QMUSC = Copper mine production capacity, United States, in 1000 tons,

PCUUS = Average annual price of copper, electrolytic copper, wirebars, U.S. producers (f.o.b. refinery), U.S. $ per ton,

QMUS = Copper mine production, United States, in 1000 tons,

UCUS = User cost of capital index, mining, United States, 1975 = 100,

D67 = Strike dummy variable, 1 in 1967, 0 otherwise,

D75/76 = Dummy variable for recession caused shifting of investments from 1975 to 1976, 1 in 1975, -1 in 1976, 0 otherwise.

Besides the equations above, Appendix III also lists the estimated equations of the mine production capacities for Canada, Chile, Peru, the Philippines, South Africa, Zaire, and for the rest of the market economies together.

With one exception, Zambia, the intercept terms of all copper mine production capacity equations are not significantly different from zero. This supports the above a priori assumption of net cash flow maximization, because it is an implication of this hypothesis as was seen above.

The coefficient of the lagged copper mine production capacity is highly significant for all countries and for the rest of the market economies together.

The geometric declining lag implicitly assumed in the estimated equations is plausible, because of the importance of short run expansions described earlier. Furthermore, regression experiments with other lag distributions also gave evidence for a Koyck lag such that in the final version of the model this approach was retained.

The Importance of Price Risk

The coefficient of the sample price variance R never reaches an asymptotic t-statistic of two in absolute value.

One explanation for the small significance is that this proxy captures the price risk inadequately. Indeed, "risk, like beauty, lies in the eyes of the beholder" and among the many possible risk measures the choice of a five year variance is somewhat arbitrary[6]. On the other hand, it is easily shown that, depending on the form and degree of relative risk aversion, increased price risk can lead to decreased or to increased output, in the latter case necessitating an increased production capacity. This case is particularly likely if risk aversion is greater at low prices, which seems to be a reasonable assumption for some of the developing countries. Therefore, economic theory does not lead to an a priori hypothesis with regard to the magnitude or sign of the coefficient of the risk variable.

Thus, I retained the risk variable in the estimated equations, because I feel that price risks influence investment decisions, although only to a small degree.

Structural Change

The period since the second World War is characterized by substantial changes in the ownership structure of the copper mines in many developing economies. It may be presumed that this transition led to a different long-run investment behavior in some of these countries, although, as we saw above, the hypothesis cannot be rejected that their short-run behavior with regard to current production remained unchanged.

Indeed, an F-test indicated structural breaks for the capacity equations of Chile, Peru and Zaire in the years of nationalization or in the beginning of a majority ownership of the host country. Dummy variables had to be introduced for this incision, because not enough data was available to estimate two separate capacity equations for each country or to use more sophisticated approaches to overcome the problem of structural change.

These structural breaks in the behavior of the developing copper producers are not surprising, however, because the neoclassical assumptions underlying this approach to the determination of the optimal capacity level are most likely too unrealistic for some of the developing countries after the takeover of the copper mining operations within their borders. Considerations with regard to employment or foreign exchange revenues also influence capacity decisions, the maximization of the net cash flow is at least not the only target of copper producers in developing countries.

In the long run other factors also determine their capacity levels. These variables cannot either be observed directly (Leibenstein's X-inefficiency for instance), or they are not yet incorporated in the current version of the model, like, for example, the level of capacity utilization. In my regression experiments, the latter variable proved to be highly significant in most cases. Proxies for the availability of foreign capital or for foreign exchange constraints were less significant, but they often had the expected sign.

To avoid "ad-hockery" (Griliches), I retained for the time being the more stringent neoclassical approach described above. Also, chapter 10 shows that dynamic simulation experiments lead to encouraging accuracy statistics for the production capacities of most of the developing copper producing economies.

7.6 Notes

1. See, for example, the results in Pobukadee's (1980) study.

2. Berndt, Morrison and Watkins (1980) elaborate on this argument in the context of energy demand models.

3. This approach follows Hall, Jorgenson (1967), Coen (1968) and Preston (1972).

4. For a more detailed derivation see Hall and Jorgenson (1967: 394-395).

5. Section 2.3 suggested an 18 months gestation period for short-term mine expansions, but the model is based on annual data. Therefore, when assuming the partial adjustment hypothesis, it is a priori equally plausible to adopt a period of one or two years as the base period of adjustment. However, regression experiments generally yielded a better explanation of historical data by postulating a one year adjustment lag only. This indicates that for some mine expansion projects the gestation period is less than the one and a half years mentioned above.

6. In many regression experiments with other variables, which possibly could have represented the price risk, I did not obtain clear-cut results. See Wagenhals (1982) for references to econometric models of agricultural supply response which include risk variables.

Chapter 8: Demand

8.1 Overview

This chapter deals with the theoretical foundations and the estimation of the demand equations.

Chapter 3 showed that the demand for copper is derived from the demand for a variety of products in many sectors of the economy. It can be demand for immediate consumption or demand for storage. Therefore, consumption equations and equations explaining the demand for storage have been derived.

The estimation of the copper consumption equations follows traditional lines in econometric copper market modeling, by and large. Copper consumption depends on the prices of the red metal and of its main substitute, both deflated with a wholesale price index as a proxy for a weighted average of the prices of other substitutes, and on an index of industrial production as a measure for the output of the copper consuming industries.

Due to the similarities of the business cycles in the main consumer countries, Zellner's method of seemingly unrelated regressions (SUR) had to be used to account for the common covariance structure of the residuals in the consumption equations. A Box-Cox procedure suggested a logarithmic functional form for the consumption equations.

Zellner's SUR approach was also used to estimate the consumption of refined copper alone (excluding the direct use of scrap). This part of the copper consumption is assumed to be just an affine function of the total copper use.

In copper market models published hitherto, inventories have often been derived as residuals in identities closing the models. In this study, they are explained by behavioural equations based on economic theory. The speculative inventory behavior of copper producers and consumers is derived from the assumption that inventory holders maximize their expected profits. For the first time in an econometric copper market model, copper futures prices are interpreted as rationally formed market anticipations based on the complete information set of the economic agents, including their expectations. In addition, the classical determinants of inventory demand due to transaction motives are also considered.

8.2 Consumption: Theory

A copper consumption function exists, provided that a copper user's marginal profitability conditions are fulfilled and that his input requirement set is locally convex in a neighborhood of the optimum. Then, the function describing the consumption of copper is a conditional factor demand function and thus depends on the relative prices of copper and its main substitute and on an index of industrial production as a proxy for the level of output produced by the copper user.

In the consumption equations specified in this econometric model, the relevant copper price is generally the world market price formed at the London Metal Exchange, converted to the currencies of the consuming countries by using spot exchange rates. The only exception is the United States, where the U.S. producers' price has been used.

The price of aluminum, copper's most important substitution product, is included in all but one of the estimated equations. For the United Kingdom the parameter of the aluminum price was insignificant and had the wrong sign. Instead, a time trend was included. The relevant aluminum price is always the London or LME price, with the exception of the United States, where the producer price is used. The German aluminum price, which is generally used in econometric copper market models, leads sometimes to more significant results in terms of the asymptotic t-statistics of the estimated coefficients, but I did not follow this approach, because the German aluminum market is not more competitive than other local aluminum markets, as is sometimes suggested, and because the published prices are not representative for arms-length prices[1].

In all equations copper and aluminum prices are deflated by a country specific wholesale price index to account for other substitution possibilities of the red metal, described in section 3.3.

To sum up: under the assumptions mentioned above the general form of a copper consumption function f is

$$q^c = f(p, p^{AL}, y),$$

where q^c = copper consumption,

p = deflated copper price,

p^{AL} = deflated aluminum price, and

y = index for the copper user's output.

Unlike in the determination of the mine production and capacity equations, economic theory did not suggest an a priori hypothesis concerning the choice of the functional form of f. Therefore I reverted to a Box-Cox procedure[2], which suggested a format logarithmic in all variables. Thus, the typical equation finally estimated reads:

$$\log q_t^c = \gamma_o + \gamma_1 \log p_{t-1} + \gamma_2 \log p_{t-1}^{AL} + \gamma_3 \log y_t,$$

where the lagged adjustment of consumption to price changes is due to the long-term contracts in the international copper trade.

8.3 Consumption: Estimation

Total Copper Consumption

Consumption equations with the format derived above were estimated simultaneously with Zellner's SUR method for total copper use in the Federal Republic of Germany, Italy, Japan, the United Kingdom, the United States, and for an aggregate consisting of all other market economies[3]. I avoided dummy variables with only one exception: the extraordinary increase in the 1966 U.S. copper consumption, which, according to trade journals at the time, can be attributed unambiguously to the preparation for the 1967/68 strike.

As a typical estimation result, the total copper consumption of the Federal Republic of Germany, is presented in

Table 8-1:

Estimated copper consumption equation, Federal Republic of Germany

$$\log(QCGE) = \underset{(7.35)}{2.71} - \underset{(-2.75)}{.0814} \log((PCULME_{-1}*REXGE_{-1})/(REXUK_{-1}*PWIGE_{-1}))$$

$$+ \underset{(3.67)}{.278} \log((PALLME_{-1}*REXGE_{-1})/(REXUK_{-1}*PWIGE_{-1}))$$

$$+ \underset{(19.8)}{.748} \log(IIPGE),$$

$$DW = 1.80,$$

where

QCGE   = Total copper consumption, Federal Republic of Germany, in 1000 tons,

PCULME = Average annual price of copper, electrolytic copper, wirebars, London Metal Exchange, cash, Pound sterling per ton,

REXGE = Exchange rate, Deutsche Mark per U.S. Dollar,

REXUK   = Exchange rate, Pound sterling per U.S. Dollar,

PWIGE   = Price index of industrial output, Federal Republic of Germany, 1975 = 1.0,

PALLME = Average annual price of aluminum, 99.5 % ingot, lb. per ton,

IIPGE   = Index of industrial production, Federal Republic of Germany, 1975 = 100.

The West German copper consumption equation is typical for the copper consumption equations of the other industrialized countries also listed in Appendix III: All variables in all copper consumption equations have the expected negative sign for the own price elasticities and the expected positive sign for the cross-price elasticities. The elasticities of the deflated prices are always significantly different from zero, but they are small compared to the income elasticities. Thus, the estimated equations indicate that the use of copper is influenced more by the general level of industrial activity than by prices. This is due to the fact that the copper input usually amounts to only a small fraction of the total materials' input in the copper using industries.

Refined Copper Consumption

Based on the assumption that the refined copper use depends on the same determinants as the total copper use and is just an affine function of this variable, I estimated equations for the consumption of refined copper only (i.e., excluding the direct use of scrap). For West Germany, Italy, Japan, the United Kingdom, the United States and for the rest of the market economies together, the coefficients of these equations were estimated simultaneously with Zellner's SUR procedure.

The equation for the refined copper consumption in the Federal Republic of Germany (QCRGE), for example, reads:

$$QCRGE = \begin{array}{c} -21.0 \\ (-.652) \end{array} + \begin{array}{c} .826\ QCGE, \\ (19.3) \end{array} \qquad DW = 1.79.$$

Appendix III shows that the corresponding equations for other important copper using countries look very much alike.

Copper Consumption in Centrally Planned Economies

Finally, a refined copper consumption equation was estimated for the centrally planned economies.

For these countries no information about the direct use of copper scrap is available. Therefore only an equation for the refined copper consumption (excluding the direct use of scrap) was estimated. It is presented in

Table 8-2:

Estimated refined copper consumption equation, centrally planned economies

$$\log(QCREW) = \quad 4.59 \quad - \quad .0155 \ \log(PCULME_{-1}/(REXUK_{-1}*PCIF_{-1}))$$
$$\qquad\qquad\quad (34.4) \qquad (-.643)$$

$$\qquad\qquad\qquad + \quad .661 \ \log(IIPEW),$$
$$\qquad\qquad\qquad\quad (25.1)$$

$$\bar{R}^2 = .963, \qquad\qquad\qquad\qquad DW = 2.04,$$

where

QCREW = Refined copper consumption (excluding direct use of scrap), centrally planned economies, in 1000 tons,

IIPEW = Index of industrial production, centrally planned economies, 1975 = 100,

PCIF = International price index, unit values of manufactures (SITC 5-8), 1975 = 100, and

PCULME and REXUK are already defined in table 8-1.

Table 8-2 shows that the refined copper consumption of the centrally planned economies depends mainly on the level of industrial activity in these countries, but also somewhat on the world copper price.

As a competitive price, the LME price is most likely a bad proxy for the accounting prices used in centrally planned economies. Furthermore, there is no information about the extent of copper substitution and the prices of potential substitutes in these countries. Therefore, the above equation is clearly mis-specified and it is not surprising that the specification error shows up in autocorrelated disturbance terms. The Cochrane-Orcutt method was thus used again to allow for first order autocorrelation.

8.4 Demand for Storage: Theory

Basically, the demand for storage can be partitioned into demand due to transaction and to speculative motives, where the latter motive results from the assumption that speculators maximize their expected profits[4].

The transaction demand for storage is modeled according to the flexible accelerator hypothesis, where the desired end-of-period inventories I_t^* depend linearly on y_t, the output of copper using industries: $I_t^* = \xi_o + \xi_1 y_t$, where ξ_o and ξ_1 are constants.

Desired stocks are only partially adjusted to actual stocks I_t,

$$I_t - I_{t-1} = \theta(I_t^* - I_{t-1}), \qquad 0 < \theta < 1,$$

such that the demand for storage due to the transaction motive is

$$I_t = \theta\xi_o + \theta\xi_1 Y_t + (1-\theta)I_{t-1}.$$

Consider now speculative stocks and let I_t^S be the level of speculative stock-holding. Given a differentiable cost function $C(I_t^S)$, a speculator's realized profit is

$$\pi_t = I_t^S(p_{t+1} - p_t) - C(I_t^S).$$

Assume that a speculator chooses end-of-period stocks I_t^S to maximize the expected utility of profits:

$$\max \quad E(U(\pi_t)).$$

Let the speculator's utility function U be non-decreasing, concave and at least twice differentiable, and let the conditional variance of prices σ_t^2 be independent of P_{t+1}^e, the conditional expected value of the price in period t+1 (given the complete information set of period t).

If we expand the utility function in a Taylor series around a fixed π_o and take expected values on both sides, we get

$$E(U(\pi_t)) = U(\pi_o) + U'(\pi_o)E(\pi_t-\pi_o) + \frac{1}{2}U''(\pi_o)E(\pi_t-\pi_o) + E(H),$$

where $E(H)$ is the expected value of the truncation remainder H, an infinite sum of higher order terms.

The first order condition for the optimal amount of speculative storage is therefore

$$U'(\pi_o)(\Delta p_{t+1}^e - C'(I_t^S)) + U''(\pi_o)I_t^S(\sigma_t^2 + \Delta p_{t+1}^{e\ 2})$$

$$- U''(\pi_o)\Delta p_{t+1}^e(C(I_t^S) + \pi_o + C'(I_t^S)I_t^S)$$

$$+ U''(\pi_o)(C(I_t^S) + \pi_o)C'(I_t^S) + \frac{d\ E(H)}{d\ I_t^S} = 0,$$

where $\Delta p_{t+1}^e := p_{t+1}^e - p_t$ is the expected price change.

If a unique solution of this equation exists, if

$$C(I_t^S) = \zeta_o + \zeta_1 I_t^S$$

is a linear approximation to the inventory holding costs in a neighborhood of the optimum, and if $d\,E(H)/dI_t^S$ is ignored on the grounds that its magnitude is likely to be small, then the optimal amount of speculative storage is approximately

$$I_t^S = \frac{1 + A(\zeta_o + \pi_o)}{A(\sigma_t^2 + (\Delta p_{t+1}^e - \zeta_1)^2)} \cdot (p_{t+1}^e - \zeta_1),$$

where A is the Arrow-Pratt coefficient of absolute risk-aversion.

Assuming $\zeta_o + \pi_o > -\frac{1}{A}$ is sufficient for

$$\frac{\partial I_t^S}{\partial \Delta p_{t+1}^e} > 0,$$

if the expected marginal revenue of the speculator exceeds the marginal costs. Then the optimal amount of speculative storage is a nonlinear function of the expected price change and the conditional variance σ_t^2. A linear approximation to the nonlinear equation derived above is justified, if σ_t^2 and the coefficient of absolute risk aversion are constant and if $(\Delta p_{t+1}^e - \zeta_1)^2$ is small compared with σ_t^2.

If we assume that this approximation holds, then the demand for storage due to the speculative motive is

$$I_t^S = \gamma_o + \gamma_1 \Delta p_{t+1}^e,$$

where γ_o and γ_1 are constants.

Naturally, inventory data does not allow a partition of copper stocks according to the underlying motives of stockholding. Therefore, a hybrid approach had to be chosen to explain the actual amount of stockholding:

$$I_t = \eta_o + \eta_1 I_{t-1} + \eta_2\, p_{t+1}^e + \eta_3 i_t + \eta_4 y_t,$$

where the η_j (j=0,..,4) are constants, and where i is a proxy for the opportunity costs of storage.

Although there is still a bone of contention among economists as to whether futures prices can be used as predictors of spot prices in general, there exists considerable evidence that the LME copper futures price may be interpreted as market anticipation of rationally acting participants. Research done by Labys, Granger (1970), Labys, Rees, Elliot (1971), Burley (1974), Goss (1981), Gupta, Mayer (1981), and Gilbert (1982) confirms this hypothesis. It was therefore assumed that the difference between the LME three months futures price and the LME spot price is proportional to the expected price change[5]. On the assumption that this rational expectations hypothesis is correct, the equation above was estimated.

8.5 Demand for Storage: Estimation

Stock equations in the format derived above were estimated for the refined copper stocks at the London Metal Exchange, in the Federal Republic of Germany, in Japan, in the United States and in all other market economies together.

An example is given in

Table 8-3:

Estimated refined copper storage equation, London Metal Exchange

$$
\begin{aligned}
\text{STLME} = \; & -.678 \; + \; .353 \; \text{STLME}_{-1} \; - \; 5.35 \; \text{INTUK} \\
& (-.0241) \quad (5.18) \qquad\qquad (-2.75) \\
& + \; 1.38 \; (\text{PCULMEF}-\text{PCULME}) \\
& \quad (2.31) \\
& + \; .728 \; \text{IIPWW} \quad + \quad 278 \; \text{DUMMY}, \\
& \quad (1.88) \qquad\qquad (7.90) \\
& \bar{R}^2 = .969, \qquad\qquad\qquad \text{DW} = 2.54,
\end{aligned}
$$

where

STLME   = Refined copper stocks, London Metal Exchange, in 1000 tons,

INTUK   = Bank rate, United Kingdom, deflated with U.K. price index of industrial production,

PCULME = Average annual price of copper, electrolytic copper, wirebars, London Metal Exchange, cash, Pound sterling per ton,

PCULMEF = Average annual price of copper, electrolytic copper, wirebars, London Metal Exchange, futures, Pound sterling per ton,

IIPWW   = Index of industrial production, market economies, 1975 = 100,

DUMMY   = Dummy variable for excessive build-up of stocks in the late 1970s.

First, this equation explaining inventories in LME warehouses as well as all other stock equations were estimated without any dummy variables. But then, most of the estimated equations did not adequately reflect the large build-up of inventories after the period of excessive speculation and during the world-wide recession in the mid 1970s. Therefore, to capture these extraordinary events, I had to introduce dummy dependent variables.

For the London Metal Exchange, the hypothesis of a partial adjustment of the inventories holds, according to the high asymptotic t-statistic of the coefficient of the lagged endogenous variable. The speed of adjustment is relatively high, because the model is based on annual data, but copper stocks are traded on a day-to-day basis.

The difference between the LME three months futures price and the LME spot price has the sign expected of a variable which indicates the anticipated price change and it is significantly different from zero.

Section 5.3 showed that events at the LME are determined by the fundamentals of the world copper market. It is therefore not surprising that the index of the market economies' aggregated industrial production performed better as an indicator of the output of the copper using industries than any national activity index.

A reasonable charge per unit of copper stored was not available, therefore it was omitted. Interest rates are used as proxies for the opportunity costs of storage. In the case of the LME storage equation the coefficient has the expected sign and is significantly different from zero.

The estimated storage equation for the LME is more or less typical for all equations explaining the level of refined copper stocks, which are listed also in Appendix III.

The flexible accelerator hypothesis of stock adjustments is confirmed by the asymptotic t-statistics of the parameters of the lagged endogenous variables in all estimated storage equations. Generally, the adjustment speed is high, again due to the fact that copper stocks are traded on a daily basis.

The difference between LME copper futures and spot price converted to local currencies has always the sign expected a priori, and the estimated parameters are always significantly different from zero. This supports the above-mentioned rational expectations hypothesis in regard to the short-term behavior of LME prices.

Interest rates do not always prove to be good proxies for the opportunity costs of storage. In the equations explaining the Japanese and the "other" market economies' copper stocks, the estimated coefficients of various interest rates always had the wrong sign. But they also were not significantly different from zero, and therefore, the interest variables were skipped. In the German storage equation, the sign of the coefficient of the interest variable coefficient was also insignificant, but correct, and so I retained the interest rate as an explanatory variable.

In all estimated equations, the transaction requirements for inventories depend on the size of the output of the copper using industries, measured by the index of industrial production, or directly on the amount of copper consumed. The signs of the estimated coefficients are always correct and they are always significantly different from zero.

In the United States, also the general price level, a dummy variable reflecting strike expectations, and the level of government stocks influence private inventory holding.

8.6 Notes

1. This was confirmed by officials of the "Vereinigte Aluminium Werke A.G.", the major producer of refined aluminum in the Federal Republic of Germany.

2. Maddala (1977: 315-316) describes the Box-Cox procedure. Although our test statistic suggested a logarithmic approach, the difference to the analogous statistic derived from a linear format is not significantly different from zero at the .1 per cent level.

3. For equations estimated according to Zellner's SUR method no coefficient of determination comparable to R^2 can be defined unambiguously. See e.g. Judge et al. (1980: 251). The squared correlation coefficients of these equations estimated by OLS were generally greater than 0.9.

4. The following approach to estimate the demand for storage is described in some more detail in Wagenhals (1981a). This paper and Labys (1973: 61-83) give many references on this subject.

5. More traditionally, the backwardation can be interpreted as a short term risk variable. For this interpretation see e.g. Hicks (1946: 138).

Chapter 9: Other Equations

9.1 Overview

This chapter deals with the specification and estimation of the remaining equations of our model.

The next section derives equations for the secondary copper supply in the most important industrialized countries: the United States, Japan, the Federal Republic of Germany, and for the rest of the market economies together. The estimation results are summarized and interpreted.

Equations explaining the most important copper prices are described subsequently: the LME cash and futures price as well as the prices of American producers and the U.S. scrap price. The estimation results for the LME cash price and the U.S. producer price are presented in detail, while the other equations are listed in Appendix III.

Finally, the chapter presents two equations explaining the East-West copper trade and U.S. refined copper production. Together with some identities, they close the model.

9.2 Secondary Supply

Independent of its source, for almost all potential uses secondary copper is practically identical with primary copper. Therefore, scrap prices are in line with the LME spot price, apart from conversion charges. The scrap supply depends on the LME price, converted to the local currencies. Only the U.S. scrap price is significant on its own.

New scrap production depends on the same decisions as the copper consumption because, as was stressed in chapter 2.5, new scrap is nothing but refuse won while fabricating copper bearing products. Thus, the production of new scrap depends mainly on the level of output in the copper using industries.

Old scrap is produced by the recycling of copper bearing outworn products. The determinants of old scrap supply resemble those influencing the mine production of copper. But an approach similar to that in chapter 7 is impossible, because information about the capacities of most old scrap suppliers is not available. Therefore, I was constrained to a traditional partial adjustment hypothesis, an only partly realistic approach, though, because the speed of adjustment on the scrap market is higher than on the market for primary copper.

Summing up these considerations, the coefficents of the following equation were estimated for the Federal Republic of Germany, Japan, the United States and for the rest of the market countries together:

$$q_t^s = \kappa_o + \kappa_1 q_{t-1}^s + \kappa_2 p_t + \kappa_3 y_t,$$

where

q_t^s denotes the secondary supply,

p_t is the deflated copper (scrap) price,

y_t is an index for the output of the copper using industries, and the parameters

κ_j (j=0,..,3) are constants.

The estimated secondary supply equation for the Federal Republic of Germany, presented in table 9-1, may serve as an example for the other scrap supply equations, which are listed in Appendix III.

Table 9-1:

Estimated secondary production equation, Federal Republic of Germany

QSGE =   30.4   + .219 QSGE$_{-1}$
          (1.42)   (1.24)

          + .439 ((PCULME*REXGE)/(REXUK*PWIGE))
             (2.25)

          + .804 IIPGE,
             (3.95)

          $\bar{R}^2 = .682,$ DW = 2.09,

where

QSGE   = Production of refined copper from scrap, Federal Republic of Germany, in 1000 tons,

PCULME = Average annual price of copper, electrolytic copper, wirebars, London Metal Exchange, cash, Pound sterling per ton,

REXGE = Exchange rate, Deutsche Mark per U.S. Dollar,

REXUK = Exchange rate, Pound sterling per U.S. Dollar,

PWIGE = Price index of industrial output, Federal Republic of Germany, 1975 = 1.0,

IIPGE = Index of industrial production, Federal Republic of Germany, 1975 = 100.

In this equation, the signs of all estimated parameters are correct, although only two of them are highly significant. Appendix III shows, that, by and large, similar statements are true for all estimated secondary supply equations.

The fit of the U.S. secondary supply equation improved considerably by introducing the stock / consumption ratio as an explanatory variable. But this approach was not successful for the other scrap supply equations.

I did not use dummy variables in the estimating equations, with only one exception, i.e. for the 1967 strike in the United States.

A comparison with the estimated mine production equations indicates that the adjusted correlation coefficients of the secondary supply equations are generally lower than the corresponding coefficients of the primary supply equations. This can be traced back to the error in the specification of the scrap supply equations, which is due to insufficient data. This has been already mentioned above.

The dynamic simulation results presented in chapter 10 however demonstrate that our model reasonably reproduces the actual historical time path of the copper scrap production.

9.3 Prices

The model includes six price equations. They explain the LME cash and three months futures price, the Canadian and Chilean price and the U.S. producers' and scrap price. All estimation results are compiled in Appendix III.

London Metal Exchange

The determination of the London Metal Exchange cash price for electrolytic copper, wirebars, follows closely the dynamic stock disequilibrium model of price adjustment, discussed and applied by Hwa (1979). In addition, dummy variables for U.S. export and price controls and for the excessive speculation against the Pound sterling in 1978/79 are included. Table 9-2 presents the estimation result.

Contrary to Hwa's result, I found no direct influence of the world copper production on the LME price. The sign of the estimated coefficient was wrong and the t-statistic quite small, so I omitted the supply variable. Otherwise, I could confirm his results.

Table 9-2:

Estimated price equation, London Metal Exchange, spot price

$$PCULME = \underset{(-3.14)}{-275} + \underset{(4.12)}{.0871\ QCWW} - \underset{(-.258)}{.0125\ STWW}_{-1}$$

$$\underset{(-2.33)}{-13.0\ T} + \underset{(7.50)}{9.96\ EUVI*REXUK} + \underset{(6.04)}{148\ DEX} - \underset{(-2.26)}{130\ D78},$$

$$\bar{R}^2 = .962, \qquad\qquad DW = 1.53,$$

where

PCULME = Average annual price of copper, electrolytic copper, wirebars, London Metal Exchange, cash, Pound sterling per ton,

QCWW = Total copper consumption, market economies, in 1000 tons,

STWW = Refined copper stocks, market economies, in 1000 tons,

T = Time trend, 1963 = 10, increasing (decreasing) by 1 each succeeding (preceding) year,

EUVI = Index of export unit values, industrialized countries, 1975=1.0,

REXUK = Exchange rate, Pound sterling per U.S. Dollar,

DEX = Dummy variable for U.S. copper export and price controls,

D78 = Dummy variable for excessive speculation against the Pound sterling.

The LME futures price is determined primarily by the same fundamentals of world demand and supply as the cash price and therefore depends on this price. As described in Chapter 4, exchange rate considerations influence hedging, speculation and arbitrage conditions on the futures market. Therefore, the futures price depends also on the Pound sterling / U.S. dollar exchange rate. The gold price reflects speculative price expectations on the precious metal markets spilling over to the copper market.

Producer Prices

The Chilean price equals the world market price, with the only exception of the early 1960s, when the producers of the Gran Minera attempted to pursue an independent pricing policy.

The specification of the North American producer price equations reflects the copper market structure in the United States and Canada. The producers can shift their production costs to a certain extent, but they cannot evade the influence of the world market price. In both countries, producers responded to the end of U.S. export controls in 1970 and to the end of price controls in 1974 in the same way: they increased prices. These events are captured by the same dummy variables in both equations.

Table 9-3 shows the estimation results for the U.S. producer price, the results for the Canadian price look very similar.

Table 9-3:

Estimated price equation, United States, producer price

$$PCUUS = \underset{(1.10)}{61.8} + \underset{(3.98)}{.283} \; PCULME/REXUK$$

$$+ \underset{(10.4)}{952} \; COSTUS + \underset{(2.04)}{211} \; D70 + \underset{(1.86)}{212} \; D74,$$

$$\bar{R}^2 = .952, \qquad\qquad DW = 1.13,$$

where

PCUUS = Average annual price of copper, electrolytic copper, wirebars, U.S. producers (f.o.b. refinery), U.S. \$ per ton,

COSTUS = Unit cost of mining index, United States, 1975 = 1.0,

D70 = Dummy variable for end of export controls,

D74 = Dummy variable for end of price controls,

PCULME and REXUK are defined in table 9-2.

Finally, in accordance with the explanations in section 4.3, the model assumes that the U.S. scrap price is determined competitively and depends on the LME futures price.

There are just a few more equations missing to complete the model.

9.4 Closing the Model

Apart from several identities listed in Appendix III, equations explaining the East-West trade in refined copper and the U.S. refined copper production close the model.

East-West Trade

To explain the East-West trade in refined copper, the model assumes that the market economies' copper net imports depend on the centrally planned countries' excess supply, the deflated LME price and the industrial activity in the market economies. Because of bureaucratic restrictions and as a consequence of long term contracts in the East-West trade, a lagged adjustment process can be ascertained. Table 9-4 shows the estimation results.

Table 9-4:

Estimated East-West trade equation

$$QIMWW = \begin{array}{c} 8.76 \\ (.175) \end{array} + \begin{array}{c} .556\ QIMWW_{-1} \\ (3.46) \end{array} + \begin{array}{c} .320\ (QMEW - QCREW) \\ (1.58) \end{array}$$

$$\begin{array}{c} -\ 2.60\ PCULME/(REXUK*PCIF) \\ (-2.83) \end{array} + \begin{array}{c} 1.57\ IIPWW, \\ (1.94) \end{array}$$

$$\bar{R}^2 = .542, \qquad\qquad DW = 2.19,$$

where

QIMWW = Net imports of refined copper from centrally planned economies to market economies,

QMEW = Copper mine production, centrally planned economies, in 1000 tons,

QCREW = Refined copper consumption, centrally planned economies, in 1000 tons,

PCIF = International price index, unit values of manufactures (SITC 5-8), 1975 = 100,

IIPWW = Index of industrial production, market economies, 1975 = 100.

PCULME and REXUK are defined in table 9-1.

In comparison with the other equations of the model, the above adjusted coefficient of determination is relatively small. To avoid this, other authors introduced dummy variables in similar equations, however, they did not succeed in justifying them with convincing economic or political arguments.

In spite of the low correlation coefficient, the historical dynamic simulation of the East-West trade equation captures the main turning points of the historical trade development reasonably well. Therefore, I dispensed with these dummy variables.

U.S. Refined Copper Production

The refined copper consumption in the United States depends on the output of U.S. copper mines currently available, on the stock / consumption ratio and on the same strike dummy, which has already been used in the U.S. mine production equation.

Table 9-5 confirms this: all estimated coefficients have the expected signs and they are highly significant.

Table 9-5:

Estimated refined copper production equation, United States

$$QRUS = \begin{array}{c} 1025 \\ (16.7) \end{array} + \begin{array}{c} .432\ QMUS \\ (8.72) \end{array} - \begin{array}{c} 1192\ STUS/QCUS \\ (-6.14) \end{array} - \begin{array}{c} 319\ DSTUS, \\ (-8.56) \end{array}$$

$$\bar{R}^2 - .906, \qquad\qquad DW = 1.57,$$

where

QRUS = Refined copper production, United States, in 1000 tons,

QMUS = Copper mine production, United States, in 1000 tons,

STUS = Refined copper stocks, United States, in 1000 tons,

QCUS = Total copper consumption, United States, in 1000 tons,

DSTUS = Dummy variable for major strikes in the U.S. copper industry.

This concludes the description of the theory and the estimation results. The next chapter deals with the dynamic properties of the model and it reports the results of some dynamic simulation experiments.

Chapter 10: Historical Dynamic Solution and Sensitivity Analysis

10.1 Overview

The last few chapters introduced a new econometric model of the world copper industry. This final chapter considers whether the above model sufficiently tracks historical copper market developments in the last three decades to be used as a tool in policy simulations and market forecasting.

The next section shows the results of a dynamic, non-stochastic simulation[1]. The model is solved simultaneously with the historical values of the exogenous variables and with the solution values generated by the model for the lagged endogenous variables[2]. Many summary goodness of fit statistics and dynamic simulation results for the model's most important aggregate variables demonstrate that the model traces the historical development of the world copper industry very well.

Then, this chapter presents the results of several multiplier simulation experiments. The third section examines one-time and sustained changes in the endogenous variables by the given changes in several of the model's exogenous variables: in aluminum prices, in industrial activity, in production costs, and in the U.S. strategic stockpile. The signs of the effects and the numerical values of the multipliers are as suggested by economic theory and by past experience in the world copper industry.

Both the dynamic simulation results and multiplier experiments therefore illustrate that the econometric model reflects the world copper market adequately, making it a useful tool for the analysis of important policy questions and for forecasting experiments.

Like any tool, this model has its limitations. The final section of this chapter points to some problems. But, taking these factors fully and properly into account, the above econometric model can serve as a useful tool in decision making. The last few paragraphs of this book refer to copper market studies based on the above model and published elsewhere, including more potential applications.

10.2 Historical Dynamic Simulation

Summary Results

For the primary and secondary supply functions, for the consumption and for the price equations, the mean absolute percentage errors of the variables are generally below 5%. For these equations, they are always below 10% with only three exceptions (Peru's production and capacity and the small copper producers' aggregate capacity). The coefficient of correlation between the actual and the simulated series is higher than 0.9 for almost all variables.

The decomposition of Theil's U-statistic is very informative. It clarifies that U_3, the non-systematic part of the simulation error, is generally high: for more than one third of the simulated variables, the fraction of the error which is due to the variance of the residuals is greater than 99%. It is greater that 70% for more than 90% of all variables. The copper stock variables, which do not perform well in terms of their mean average percentage error, show a rather high disturbance proportion U_3. This indicates that the simulation error for the stockholding equations is largely non-systematic.

Table 10-1 shows summary goodness of fit measures. The rest of the section adds detailed information about the undisturbed historical dynamic solution values of some of the model's most important aggregate endogenous variables.

Table 10-1:

Accuracy Statistics of the Historical Dynamic Solution, 1956-1980

Explanation of Symbols:

R   = Correlation coefficient of actual and simulated series,

RMSE = Root-mean-squared error,

MAE = Mean absolute error,

MAPE = Mean absolute percentage error,

C   = Regression coefficient of actual on simulated series,

U   = Theil's inequality coefficient,

U_1 = Fraction of error due to bias,

U_2 = Fraction of error due to difference of regression coefficient from unity,

U_3 = Fraction of error due to residual variance.

Table 10-1 (continued)

Variable	R	RMSE	MAE	MAPE	C	U	U_1	U_2	U_3
PCUCA	.986	4.01	3.12	6.91	1.00	.0346	.0038	.0000	.9962
PCUCH	.980	93.7	80.0	8.89	1.02	.0379	.0010	.0102	.9888
PCULME	.986	39.4	34.3	9.30	1.01	.0356	.0037	.0040	.9923
PCULMEF	.987	37.9	32.8	8.75	1.01	.0343	.0034	.0023	.9943
PCUUS	.976	97.4	73.6	7.80	1.01	.0414	.0046	.0010	.9944
PSUS	.958	81.1	63.1	8.96	1.04	.0494	.0000	.0174	.9826
QCGE	.969	37.2	31.4	2.14	.984	.0244	.0007	.0038	.9955
QCIT	.993	14.8	12.0	4.19	.996	.0186	.0000	.0015	.9985
QCJA	.992	59.1	46.8	3.83	.989	.0275	.0002	.0072	.9926
QCOM	.988	92.1	73.5	5.26	.982	.0220	.0020	.0013	.9850
QCREW	.998	35.7	25.0	3.81	.971	.0118	.0734	.2007	.7259
QCRGE	.950	40.4	31.7	5.42	1.01	.0331	.0007	.0008	.9985
QCRIT	.980	14.6	11.5	5.36	.998	.0287	.0014	.0001	.9985
QCRJA	.991	52.1	42.3	7.41	.993	.0333	.0000	.0023	.9977
QCROM	.986	81.5	62.3	4.20	.989	.0251	.0008	.0042	.9948
QCRTW	.996	164.	129.	2.03	.983	.0118	.0147	.0359	.9494
QCRUK	.843	28.4	21.6	3.95	1.21	.0269	.0142	.0657	.9201
QCRUS	.902	141.	115.	7.06	1.01	.0403	.0004	.0003	.9993
QCRWW	.993	161.	123.	2.40	.999	.0147	.0040	.0070	.9990
QCUK	.864	31.2	25.5	3.68	1.14	.0232	.0114	.0448	.9438
QCUS	.884	199.	157.	6.55	1.00	.0387	.0000	.0000	1.000
QCWW	.987	267.	185.	2.68	.986	.0178	.0012	.0075	.9913
QMCA	.982	30.8	23.7	4.29	.981	.0266	.0375	.0099	.9526
QMCAC	.980	41.3	34.5	5.38	.978	.0304	.0324	.0118	.9558
QMCH	.955	61.3	50.6	7.01	.991	.0408	.1303	.0008	.8689
QMCHC	.978	44.2	32.5	4.32	.964	.0275	.2558	.0219	.7223
QMEW	.998	30.9	25.1	2.32	1.01	.0128	.0118	.0098	.9784
QMOM	.938	93.6	74.4	9.95	1.05	.0588	.0554	.0148	.9298
QMOMC	.969	123.	93.4	11.5	.873	.0658	.4209	.1427	.4364
QMPE	.904	40.7	33.6	22.2	1.03	.0917	.0759	.0044	.9197
QMPEC	.951	32.4	24.8	15.1	1.06	.0659	.1318	.0244	.8438
QMPH	.992	12.2	9.33	8.87	.982	.0358	.0249	.0195	.9556
QMPHC	.991	14.5	11.0	9.34	1.01	.0382	.0119	.0017	.9864
QMSA	.978	20.4	15.9	10.2	.883	.0582	.3507	.1791	.4702
QMSAC	.986	19.6	14.6	8.51	.881	.0543	.4285	.2242	.3474
QMTW	.993	212.	171.	3.17	.978	.0175	.1740	.0277	.7982
QMUS	.969	56.9	45.5	3.82	.941	.0232	.0223	.0556	.9221
QMUSC	.991	43.7	34.7	2.50	.919	.0146	.0640	.2878	.6482
QMWW	.989	202.	162.	3.69	.964	.0207	.2077	.0426	.7497
QMWWC	.991	269.	219.	4.14	.931	.0241	.3455	.1462	.5083
QMZI	.947	32.1	25.4	10.3	.880	.0426	.1488	.1197	.7316
QMZIC	.950	38.4	29.4	7.77	.890	.0456	.1044	.1044	.7378
QMZM	.924	36.2	28.2	9.78	.903	.0291	.0140	.0618	.9242
QMZMC	.987	19.1	13.8	2.25	.941	.0140	.0331	.1198	.8471
QRUS	.971	38.5	30.5	2.20	1.01	.0137	.0045	.0026	.9929
QSGE	.866	15.0	12.7	8.96	1.11	.0487	.0120	.0283	.9597
QSJA	.926	12.0	9.61	14.3	.978	.0701	.0002	.0032	.9966
QSOM	.920	31.4	26.7	9.07	1.03	.0501	.0064	.0040	.9896
QSUS	.973	22.6	17.7	5.08	1.02	.0313	.0086	.0107	.9807
QSWW	.975	49.2	39.8	4.49	1.02	.0270	.0014	.0064	.9922
STGE	.945	9.95	7.66	12.7	1.02	.0657	.0012	.0029	.9959
STJA	.895	45.8	29.8	48.5	1.16	.1850	.0035	.0683	.9282
STLME	.987	31.3	22.1	79.7	1.03	.0704	.0019	.0273	.9708
STOM	.983	21.8	17.1	19.4	1.01	.0586	.0307	.0025	.9668
STUS	.977	30.0	23.1	14.7	1.01	.0569	.0093	.0017	.9890
STWW	.991	81.6	65.2	14.8	1.07	.0488	.0012	.1941	.8047

Detailed Results

Above, I presented summary goodness of fit results for all endogenous variables. Detailed listings and graphs of all dynamic simulation results are beyond the scope of this study. Due to space restrictions this presentation has been restricted to a few salient variables.

The following tables and figures present and show the time paths of the actually observed values ("*") and of their dynamic simulation values ("x") for several of the most important aggregate variables of the world copper model.

In the figures below, the data points are joined by cubic splines. Straight lines connect the actual data points. Dotted lines link the historical dynamic simulation values generated by the model. The horizontal is the time axis. The following variables and their dynamic simulation values are plotted:

QMTW   = Copper mine production, total world, in 1000 tons,

QCRTW = Refined copper consumption, total world, in 1000 tons,

QCWW   = Total copper consumption, market economies, in 1000 tons,

QSWW   = Production of refined copper from scrap, market economies, in 1000 tons,

STWW   = Refined copper stocks, market economies, in 1000 tons,

PCULME = Average annual price of copper, electrolytic copper, wirebars, London Metal Exchange, cash, Pound sterling per ton,

PCUUS = Average annual price of copper, electrolytic copper, wirebars, U.S. producers (f.o.b. refinery), U.S. $ per ton.

The tables and graphs demonstrate show that the model captures most of the turning points in the historical development of the world copper market since the mid 1950s. They confirm that the development of the world copper market is described reasonably and sufficiently by the current model.

It was therefore used to perform a sensitivity analysis, which will be described in the next section.

Dynamic simulation: Copper mine production, total world (QMTW)

Figure 10-1:

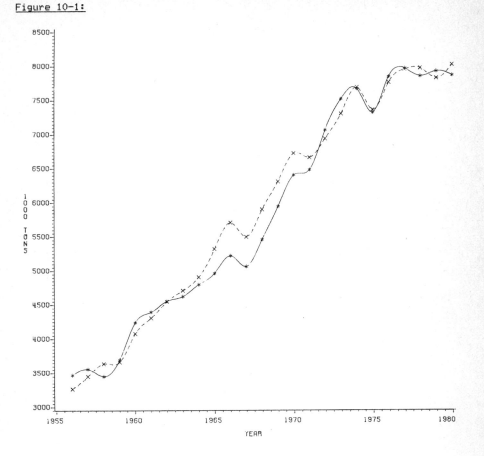

Table 10-2:

Year	Actual value	Solution value	Year	Actual value	Solution value
1956	3470	3267			
1957	3556	3448	1969	5943	6309
1958	3449	3638	1970	6403	6723
1959	3693	3661	1971	6477	6663
1960	4242	4080	1972	7063	6938
1961	4394	4306	1973	7514	7300
1962	4555	4548	1974	7670	7690
1963	4624	4710	1975	7317	7357
1964	4799	4906	1976	7843	7763
1965	4963	5318	1977	7965	7952
1966	5220	5706	1978	7846	7969
1967	5060	5495	1979	7925	7823
1968	5459	5900	1980	7862	8019

Dynamic simulation: Refined copper consumption, total world (QCRTW)

Figure 10-2:

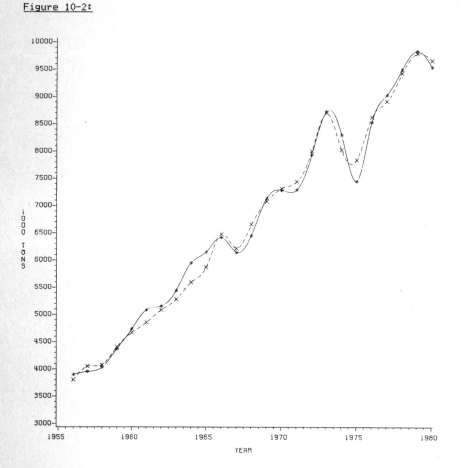

Table 10-3:

Year	Actual value	Solution value	Year	Actual value	Solution value
1956	3898	3801			
1957	3955	4050	1969	7137	7084
1958	4033	4079	1970	7291	7320
1959	4374	4408	1971	7295	7447
1960	4736	4672	1972	7942	8009
1961	5088	4856	1973	8740	8706
1962	5160	5089	1974	8306	8029
1963	5448	5283	1975	7452	7848
1964	5953	5597	1976	8541	8628
1965	6152	5875	1977	9041	8921
1966	6421	6483	1978	9502	9438
1967	6147	6217	1979	9845	9797
1968	6457	6665	1980	9546	9663

Dynamic simulation: Total copper consumption, market economies (QCWW)

Figure 10-3:

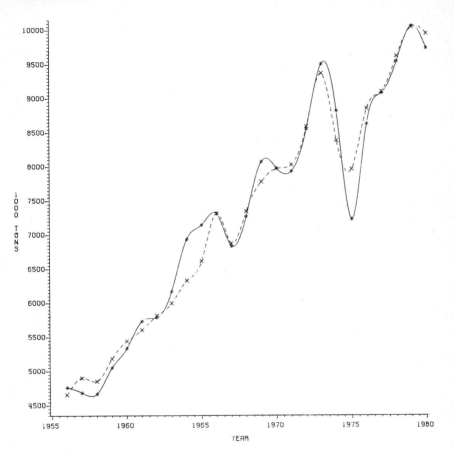

Table 10-4

Year	Actual value	Solution value		Year	Actual value	Solution value
1956	4763	4654				
1957	4682	4901		1969	8070	7778
1958	4666	4849		1970	7975	7967
1959	5047	5184		1971	7927	8024
1960	5335	5441		1972	8549	8585
1961	5732	5602		1973	9504	9366
1962	5785	5815		1974	8817	8382
1963	6170	5996		1975	7222	7960
1964	6933	6327		1976	8624	8848
1965	7141	6611		1977	9073	9097
1966	7309	7315		1978	9542	9615
1967	6833	6873		1979	10057	10040
1968	7271	7346		1980	9731	9945

Dynamic simulation: Production of refined copper from scrap, market economies (QSWW)

Figure 10-4:

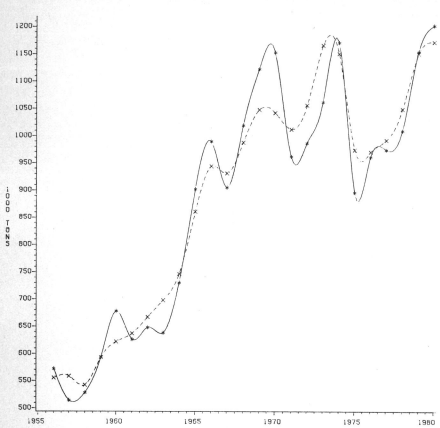

Table 10-5:

Year	Actual value	Solution value	Year	Actual value	Solution value
1956	572	555			
1957	514	558	1969	1123	1048
1958	528	542	1970	1153	1042
1959	593	593	1971	962	1012
1960	678	622	1972	987	1056
1961	626	637	1973	1062	1166
1962	648	667	1974	1171	1151
1963	638	698	1975	896	974
1964	730	746	1976	961	970
1965	902	861	1977	974	992
1966	990	944	1978	1059	1049
1967	905	931	1979	1188	1151
1968	1019	988	1980	1202	1172

Dynamic simulation: Refined copper stocks, market economies (STWW)

Figure 10-5:

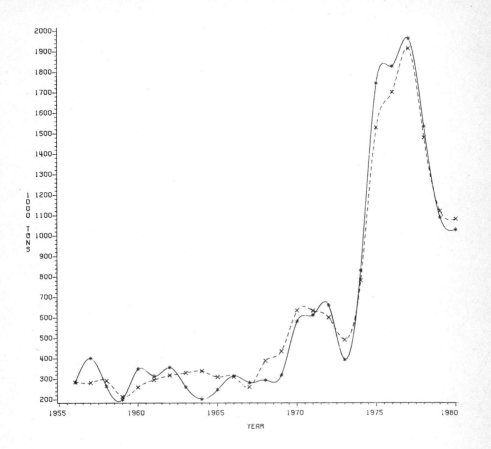

Table 10-6:

Year	Actual value	Solution value		Year	Actual value	Solution value
1956	285	285				
1957	402	282		1969	319	436
1958	265	291		1970	582	634
1959	200	214		1971	612	633
1960	349	260		1972	659	599
1961	314	296		1973	394	492
1962	356	318		1974	832	781
1963	261	330		1975	1744	1529
1964	202	338		1976	1828	1701
1965	248	309		1977	1964	1914
1966	315	311		1978	1535	1479
1967	282	261		1979	1080	1121
1968	294	388		1980	1028	1082

Dynamic simulation: Average annual price of copper, electrolytic copper, wirebars, London Metal Exchange, cash (PCULME)

Figure 10-6:

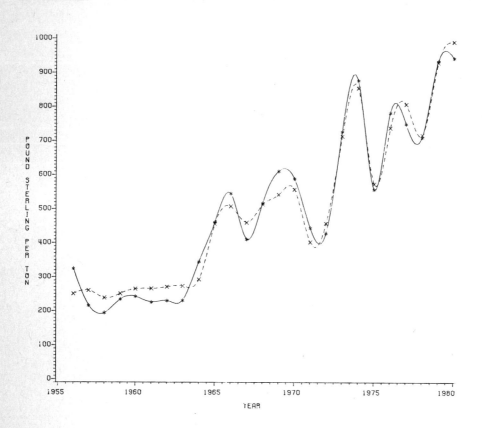

Table 10-7:

Year	Actual value	Solution value	Year	Actual value	Solution value
1956	324	250			
1957	216	260	1969	611	542
1958	194	238	1970	589	557
1959	234	251	1971	444	402
1960	242	265	1972	428	457
1961	226	266	1973	727	714
1962	230	271	1974	878	855
1963	231	273	1975	557	573
1964	345	293	1976	781	738
1965	461	456	1977	751	807
1966	546	508	1978	710	717
1967	411	460	1979	936	930
1968	517	514	1980	941	990

Dynamic simulation: Average annual price of copper, electrolytic copper, wirebars,
U.S. producers (PCUUS)

Figure 10-7:

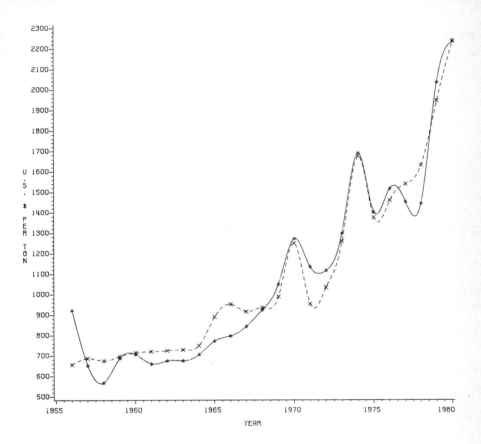

Table 10-8:

Year	Actual value	Solution value	Year	Actual value	Solution value
1956	922	657			
1957	652	687	1969	1048	988
1958	568	675	1970	1272	1250
1959	687	697	1971	1134	953
1960	707	715	1972	1116	1034
1961	660	720	1973	1298	1258
1962	675	725	1974	1690	1675
1963	675	729	1975	1401	1374
1964	705	750	1976	1517	1459
1965	772	890	1977	1451	1538
1966	797	951	1978	1444	1632
1967	843	915	1979	2036	1946
1968	923	935	1980	2236	2238

10.3 Multiplier Simulation Experiments

This section reports the results of eight simulation experiments, which reflect the dynamic response properties of the econometric model of the world copper market.

The eight experiments, broken in four categories:

1. A 10% increase of all aluminum prices:
    a) a sustained increase,
    b) a one-time increase in 1960.

2. A 5% increase of the industrial production in all market economies:
    a) a sustained increase,
    b) a one-time increase in 1960.

3. A 10% increase in the production costs and in the user cost of capital of all copper producing countries:
    a) a sustained increase,
    b) a one-time increase in 1960.

4. An increase in the U.S. government's strategic stockpile by 100,000 tons:
    a) a sustained increase,
    b) a one-time increase in 1960.

Results

Experiment 1 a

A sustained increase in the aluminum prices, beginning in 1956, leads to a substitution of copper for aluminum. U.S. consumption increases by 30,000 tons, the market countries' aggregated consumption by 100,000 tons. This raises the LME price of copper, and, to a smaller extent, the two North American prices. This leads to an increased mine production and additional capacities in all copper producing countries. Copper production increases only slightly less than the refined consumption, such that the Western World's total private refined copper stocks increase by some 5,000 tons in the long run.

Experiment 1 b

A one time 10% increase of the aluminum prices in 1960 leads to a short run substitution of copper for aluminum. In 1961, the -market economies' total copper consumption increases almost 80,000 tons. Simultaneously the LME price grows some 3%. In the following year, the rise of the copper price / aluminum price ratio leads to a short run substitution back to aluminum. At the same time, mine capacities and production grow due to the high copper prices. In feedback, both factors tend to depress prices, which in 1962 do not quite reach the levels of the control solution. From 1963 onward, however, the prices of the control solution and the disturbed solution are identical again.

Reflecting the changes in production and consumption, total refined copper stocks decrease by 3,000 tons in 1961. One year later they remain only 800 tons above the old level, which is reached again in 1963.

Experiment 2 a

A 5% sustained increase of industrial production in all copper consuming countries increases the market countries' use of the red metal by almost 300,000 tons in the long run. In the United States alone more than 50,000 tons of copper and copper scrap are additionally consumed. This consumption growth raises copper prices. The LME price, for instance, increases by some 20 Pound sterling per ton compared to the control solution. The price change leads to capacity expansions and increased production. In the long run, the world's copper mine production and capacities reach a new level some 4% higher than in the control solution. The level of refined copper stocks remains almost constant.

Experiment 2 b

A one-time 5% increase in the industrial production of all countries in 1960 increases the market countries' total copper consumption in this year by almost 400,000 tons (94,000 tons in the United States alone). This raises the LME price more than 10% in the short run. Yet, the U.S. producer prices increases only by 4%. The price spur induces mine production and capacity expansions in all copper producing countries, which counterbalance the price increase in the next period. Due to the lagged adjustment on the primary supply side, output and capacity decline slower than consumption to the control level. Thus, for a few years copper stocks are relatively higher. In the mid 1960s all variables reach their original levels again.

Experiment 3 a

A 10% sustained increase in costs (unit production costs and unit user costs of capital) leads to a long run reduction of copper mine capacity and production. In 1980, for example, the total world's copper mine production is some 5% smaller compared to the control solution. This tends to increase copper prices and therefore supports a substitution away from copper. So simultaneously with the reduction of the copper output, the copper consumption decreases too. The LME copper price and refined copper stocks remain almost on the level of the control solution. However, the U.S. and the Canadian producer prices somewhat increase, reflecting the loosely oligopolistic structure of the North American market, where the producers are able to shift some of their cost increases directly to the consumers.

Experiment 3 b

A one-time 10% increase in the copper production and capital costs in 1960 leads to short run production cutbacks of some 25,000 tons in the total world. Copper prices increase in the United States by 6%, less at the London Metal Exchange. Due to these price increases, in the next period copper consumption declines somewhat, such that prices and stocks return to the control solution level, which is reached by all variables in the next year.

Experiment 4 a

A sustained 100,000 ton increase in the U.S. government's strategic stockpile reduces the amount of copper which can be used in the private sector. Copper prices, production, and capacity increase only slightly. Copper consumption decreases somewhat, but the change is only about 1% for almost all variables.

The main effect of the government's stock increase is a substitution of private for public storage[4]. In the United States, the level of refined copper stocks remains 18,000 tons below the control solution level, which in 1980, corresponds to some 6% of the United States' refined copper stocks in the private sector. The stocks in the rest of the world only change slightly in the long run.

Experiment 4 b

A one-time 100,000 ton increase in U.S. government stocks in 1960 leads only to small short run increases in copper prices and to corresponding changes in quantities. The main effect is a decline of private refined copper stocks in the United States by 14,000 tons in 1960. In the next year the U.S. stocks are only 3,000 tons below the control solution level, which is finally reached again in 1962.

Summary

The dynamic simulation results and multiplier experiments presented in the last two sections suggest that the above econometric copper market model tracks sufficiently history to use it for projections regarding production, capacity, consumption, prices, and inventories in the 1980s and forthcoming years. It can then be used in attempts to answer basic questions facing the copper industry at the present time.

Although the model appears to be a reasonable tool for forecasting and ana- lysis of policy questions, of course, it has its limitations.

10.4 Concluding Remarks

Open Problems

Like any econometric model, the copper market model presented in this book is based on restrictive assumptions, suffers from missing data, and works with estimators whose exact sampling distributions are unknown due to pretesting.

Although I have always tried to set forth the underlying suppositions clearly, I would like to stress three points which I think are especially problematical:

1. The assumption of a Cobb-Douglas production function implicating a unit elasticity of substitution and neglecting shifts in the composition of "the" variable factor.

2. The assumption that after the nationalization or the acquisition of a 51% ownership of the copper mines, the developing copper producers still determine their optimal capacity by maximizing their discounted net cash flow.

3. The determination of the interaction of spot and futures prices. Here economic theory is used only as a grab-bag, lacking a neat and comprehensive theoretical derivation of the interaction.

While the first two points can be subsumed under the heading "missing data", the last point results from "missing theory".

These points deserve more consideration in the near future. Naturally many other interesting problems exist. These include, for example, the modeling of the influence of uncertainty and alternative expectations schemes on the supply and demand decisions[5].

Outlook

The description and analysis of the world copper market and its institutional and economic peculiarities presented jointly with the econometric model in this book provide a framework for forecasting exercises and policy evaluations.

This model has been used for these purposes. Detailed forecasting results have been published elsewhere (Wagenhals (1984b)). Policy studies performed with the model include, for example, the assessment of the impact of alternative CIPEC policy strategies on the world copper market (Wagenhals (1984d, e)) and the evaluation of quantitative effects of potential copper mining from ocean floor nodules[6].

Apart from the subjects of the studies mentioned, other topics might be worth investigating using this econometric model. Two such applications would be: the assessment of a copper price stabilization scheme and the linkage of the commodity market model with econometric country models.

Although the discussion about the costs and benefits of a copper price stabilization scheme by means of an international buffer stock scheme has become quieter in the last few years, this topic is still a bone of contention between some copper exporting and importing countries[7]. Theoretical studies[8] have presented contradictory results, such that an empirical analysis of the problem is indispensable.

Contrary to other econometric copper market models, the model presented in this report exhibits some new features, which are useful in evaluating the outcome of copper price stablization. The most important of these features are:

1. The derivation of the mine production capacities, which allows the seizure of the capacity effects of buffer stock induced price changes on a high level of disaggregation,

2. the introduction of the futures market, which can considerably alter the results of a price stabilization scheme[9],

3.   the modeling of inventory demand equations, which permit the estimation of the
     extent of substitution of public for private storage due to a   price
     stabilization scheme.

Thus, the impact of an international buffer stock scheme on the world copper
industry is an interesting subject which can be assessed with the current model. It
may lead to new interesting results in comparison to the results of existing studies
of this topic, which neglected these factors.

In connection with econometric models of the most important developing copper
exporters, this model allows the examination of the repercussions of the world
copper market on those countries highly dependent on copper export earnings. For
example, the consequences of alternative tax systems on investments could be
evaluated. Although copper market models already have been combined with models of
the local economies in Chile and Zambia, the inclusion of such important copper
producers, such as Peru and Zaire is still missing.

Another example for a possible application of this model is the linkage with an
econometric model of an important industrialized copper consuming country. In this
context, the impact of various forms of supply restrictions by the main import
sources could be assessed either in particular, for example in respect to
non-ferrous metal fabricators, or, in general in regard to GDP or GNP.

It is even possible to estimate bridge equations which link the variables of
the copper model to a single consumer, say, a fabricator of copper and copper-alloy
semimanufactures. This allows, for instance, the examination of the impacts of
changing copper prices on the output of semis. It renders possible the assessment of
the influence of changes in the input factor mix, including copper. If desired,
bridge equations can be developed with a high degree of disaggregation, for
instance, including brass rod, strip and sheet, etc.

Thus, the econometric model developed above serves as a potential starting
point to evaluate the influences of the world copper market on certain sectors of
the economy, such as the electrical and electronics or construction industry, or
even on particular medium sized firms, like brass mills.

A range of further possible uses of econometric non-ferrous metal market models
in general is presented in Wagenhals (1983a). Certainly the few examples in this
section indicate that scope for further research exists using this econometric
model.

Last, but not least, aside from the presentation of a new econometric world copper market model with these interesting actual and potential applications, I hope that my study may contribute to the understanding of the world copper market in general, that it may help decision makers in international organizations, business and governments and that it may stimulate further research in this area.

10.5 Notes

1. I also performed a static (i.e. one-period) simulation with the model. It was solved using actual values of the lagged endogenous variables instead of using their generated solution values. Naturally, then the goodness of fit statistics were better compared to the dynamic solution of the model. But "a dynamic solution is clearly a more stringent test of a model and is clearly the exercise most like forecasting" (Klein, Young (1980: 65)). Thus, the results of the one-period simulation are not reported here.

2. For the estimation and simulation of the model, I used the TSP software package, version 3.5.

3. For a definition, explanation and critique of the goodness of fit statistics used, see, for example, Maddala (1977: 343-347)).

4. See Wagenhals (1981a) for a different approach to the substitution of public for private storage. The results presented here complete the discussion for this paper, where only an upper bound for the extent of substitution was derived.

5. Wagenhals (1984a) describes uses and limitations of econometric minerals market models in general.

6. The Kiel Institute of World Economics published several studies dealing with the impact of deep sea nodule mining on the world markets for cobalt, manganese and nickel. See Gupta (1981), Rafati (1982a, b, c), Foders, Kim (1982, 1983). In the context of this research program, the current model has been used to assess the influence of ocean floor mining on the world copper market (Wagenhals (1984b)).

7. For the development of the discussion of this topic see Billerbeck (1975), Solveen (1977) and Obernolte (1980).

8. Since the classical studies on price stabilization by Waugh (1944) and Oi (1961) there has been a surge of published and unpublished results. The best contribution to the theoretical debate is Newbery and Stiglitz (1981). Unfortunately, the bibliography of this book is not complete. For additional references see e.g. Wilson (1977), Turnovsky (1978), Adams, Behrman (1982) or Wagenhals (1982).

9. Only a few theoretical studies deal with this topic, for instance, McKinnon (1967), Turnovsky (1979) or Newbery, Stiglitz (1981: 177-192).

Appendix I: General Remarks on Methodology and Data

Methodology

The model was estimated based on annual data from 1956 to 1980.

All consumption equations were estimated simultaneously with Zellner's method of seemingly unrelated regressions (Zellner, 1962). The ordinary least squares (OLS) method and the Cochrane-Orcutt technique were used in the estimation of the other equations.

Despite its well-known deficiencies, the OLS procedure is employed very often in large-scale simultaneous econometric models. First, it is robust against specification errors and, second, "predictions from equations estimated by OLS often compare favorably with those obtained from equations estimated by the simultaneous-equation methods" (Maddala (1977: 231)). A discussion of its pros and cons is beyond the scope of this study, but see e.g. Mariano (1978: 81-84) or Klein and Young (1980: 60-61) for a critical endorsement of this approach. Because of the lengthy list of exogenous variables a full two stage least squares or maximum likelihood estimation of the unknown parameters was not feasible.

If Geary's non-parametric test or the Durbin-Watson statistic indicated first order autocorrelation of the error terms, then the Cochrane-Orcutt iterative technique was applied to reestimate the equations.

A systematic pattern in the OLS-residuals of the consumption equations revealed that the residuals were contemporaneously correlated across the copper consuming countries. This is probably due to the transmission mechanism between business cycle fluctuations in the industrialized countries. Therefore the ordinary least squares method was clearly not feasible. I used Zellner's method of seemingly unrelated regressions to estimate the unknown coefficients.

Balancing Surplus

Copper statistics are not always as satisfactory as an econometrician would wish. This is not only true for statistics on copper stocks and capacities, but also for production, consumption and scrap statistics.

The lack of consistency in stock statistics can be due to changes in transit stocks, to the volatility of scrap storage, to the revision of the basis upon which copper stock statistics are gathered, and due to random recording inaccuracies. An article published by the Commodities Research Unit (1978) excellently surveys these problems.

Not only stock statistics, but also production, consumption, scrap and capacity figures may be affected by inexactness.

Due to the spread of continuous casting the line between production and consumption of refined copper begins to blur. There are classification problems for leach cathodes or wire rods, for example. Also, the use of copper for ordnance and other military uses is not always adequately reflected by the published statistics. Changes in the relation of internal and external scrap recycling by copper using companies affect the magnitude of publicly reported scrap data.

Considering these problems, it is not surprising that balancing surpluses (or deficits) occur in copper statistics. Therefore, I defined two variables for the unaccounted balancing surplus of the United States (BSUS) and of the rest of the market economies together (BSOM) as follows:

> Recorded Production
> − Recorded Consumption
> + Recorded net imports
>
> = Calculated increase in stocks
> − Recorded increase in stocks
>
> = Unaccounted balancing surplus.

As usual in econometric copper market models, these two variables are treated as given exogenously. I suppose that both variables mainly include invisible stock changes, because they are significantly correlated with the respective apparent stock changes. Therefore one could argue that they should be explained endogenously.

Unit Costs of Mining

Although for some points in time information about mining costs is available, comparable complete time series on mining costs are not obtainable.

Therefore I constructed indices for the unit costs of mining according to the following fomulae (the labels are defined in Appendix II):

COSTCA = .8 PWICA   + .2 PWICA1
COSTCH = .4 PWICH   + .3 PWICH1 + .3 EUVI*REXCH/4.91
COSTPE = .4 PWIPE   + .6 (EUVI + EUVO)*REXPE/81.6
COSTPH = .4 PWIPH   + .6 (EUVI + EUVO)*REXPH/14.5
COSTSA = .8 PWISA   + .2 EUVI*REXSA/.732
COSTUS = .05 PWIUSE + .29 PWIUSL + .11 PWIUSM + .35 PWIUS + .20 UCUS
COSTZI = .4 PWIZI   + .6 (EUVI + EUVO)*REXZI/.643
COSTZM = .4 PWIZM   + .6 (EUVI + EUVO)*REXZM/1.29.

The weights for the United States are calculated from estimated annual operating costs of mining operations as reported in Bennett et al. (1973).

For the developing countries, the cost structure of copper mining companies is different. Production especially is less capital intensive, as the data on the cost structure of Peruvian copper mines compiled by Brundenius (1972) shows. The a priori weights defined above are plausible according to this and other sources.

As an industrialized country, South Africa depends less on imports of inputs in copper mining than the developing countries, where material, equipment and energy are often imported. This is reflected in the relative weights of international price indices, which are .2 for South Africa and .6 for the developing copper producers.

User Costs of Capital

The user costs of capital are calculated according to the formulae given in section 7.4.

For the United States a time series of the user costs of capital for mining equipment was available from the Wharton Econometric Forecasting Associates. For all other countries I calculated it, assuming straight line depreciation.

Proxies for the purchase price of a unit of capital services are given by the respective price indices, which are listed in Appendix II for all countries. For the aggregate "other market economies" the index of export unit values of the industrial countries (EUVI) is used.

The interest rate used in calculating the user costs is generally the London interbank offering rate converted to the local currencies. This reflects the international character of the financing of copper ventures in general, and particularly the existence of - at best - rudimentary markets for capital and capital goods as well as the government control of interest rates, which is typical for many developing copper producers. Apart from the United States, only for Canada is a domestic bank rate used.

The main source of information on corporate tax rates is the Price Waterhouse Information Guide on corporate taxes, including special issues for the copper producing countries, if available. For the early years of the observation period information has also been extracted from de Patiño (1965), Coleman (1975), Faber, Potter (1971), Obidegwu (1980) and from company statements. Missing observations were estimated. For the Philippines I assume that the current highest corporate tax rate for domestic corporations has always been applicable. For the aggregate "other market economies" I use a constant tax rate of 50%.

Eventually, the depreciation charge for tax purposes is estimated to be 20%, unless other information is available from the Price Waterhouse Guide. The lifetime for tax purposes is assumed to be 20 years.

Appendix II: List of Variables

Sources

BLS1   Wholesale Prices and Price Indexes, U.S. Bureau of Labor Statistics,

BLS2   Employment and Earnings, U.S. Bureau of Labor Statistics,

CBS   General Review of the Mineral Industries, Canadian Bureau of Statistics,

CDA   Copper Supply and Consumption, Copper Development Association,

CSO   Digest of Statistics, Central Statistical Office, Lusaka,

FP   Metal Statistics, Fairchild Publications,

IMF   International Financial Statistics, International Monetary Fund,

MG   Metal Statistics, Metallgesellschaft AG,

PD   Phelps Dodge Co.,

TR   Transformation (see Appendix I),

UN   Monthly Bulletin of Statistics and Statistical Yearbook, United Nations,

WBMS   World Metal Statistics, World Bureau of Metal Statistics,

WEFA   Wharton Econometric Forecasting Associates,

WB   Commodity Trade and Price Trends, World Bank,

WB79   World Bank (1979).

Endogenous Variables

Variable	Description	Source
PCUCA	Average annual price of copper, electrolytic copper, wirebars, Canadian producers, Canadian cents per lb.,	FP
PCUCH	Average annual price of copper, electrolytic copper, wirebars, Chile (c.i.f.), U.S. $ per ton,	MG, WB79
PCULME	Average annual price of copper, electrolytic copper, wirebars, London Metal Exchange, cash, Pound sterling per ton,	MG
PCULMEF	Average annual price of copper, electrolytic copper, wirebars, London Metal Exchange, futures, Pound sterling per ton,	MG

PCUUS	Average annual price of copper, electrolytic copper, wirebars, U.S. producers (f.o.b. refinery), U.S. $ per ton,	MG
PSUS	Average annual price of dealers' No. 2 heavy copper scrap, U.S. $ per ton,	MG
QCGE	Total copper consumption, Federal Republic of Germany, in 1000 tons,	MG
QCIT	Total copper consumption, Italy, in 1000 tons,	MG
QCJA	Total copper consumption, Japan, in 1000 tons,	MG
QCOM	Total copper consumption, other market economies, in 1000 tons,	MG
QCREW	Refined copper consumption, centrally planned economies, in 1000 tons,	MG
QCRGE	Refined copper consumption, Federal Republic of Germany, in 1000 tons,	MG
QCRIT	Refined copper consumption, Italy, in 1000 tons,	MG
QCRJA	Refined copper consumption, Japan, in 1000 tons,	MG
QCROM	Refined copper consumption, other market economies, in 1000 tons,	MG
QCRTW	Refined copper consumption, total world, in 1000 tons,	MG
QCRUK	Refined copper consumption, United Kingdom, in 1000 tons,	MG
QCRUS	Refined copper consumption, United States, in 1000 tons,	MG
QCRWW	Refined copper consumption, market economies, in 1000 tons,	MG

QCUK	Total copper consumption, United Kingdom, in 1000 tons,	MG
QCUS	Total copper consumption, United States, in 1000 tons,	MG
QCWW	Total copper consumption, market economies, in 1000 tons,	MG
QIMRUS	Net imports of refined copper, United States, in 1000 tons,	CDA
QIMUS	Net imports of copper, United States, in 1000 tons,	CDA
QIMWW	Net imports of refined copper from centrally planned economies to market economies,	WBMS
QMCA	Copper mine production, Canada, in 1000 tons,	MG
QMCAC	Copper mine production capacity, Canada, in 1000 tons,	PD
QMCH	Copper mine production, Chile, in 1000 tons,	MG
QMCHC	Copper mine production capacity, Chile, in 1000 tons,	PD
QMEW	Copper mine production, centrally planned economies, in 1000 tons,	MG
QMOM	Copper mine production, other market economies, in 1000 tons,	MG
QMOMC	Copper mine production capacity, other market economies, in 1000 tons,	PD
QMPE	Copper mine production, Peru, in 1000 tons,	MG
QMPEC	Copper mine production capacity, Peru, in 1000 tons,	PD
QMPH	Copper mine production, Philippines, in 1000 tons,	MG

QMPHC	Copper mine production capacity, Philippines, in 1000 tons,	PD
QMSA	Copper mine production, South Africa and Namibia, in 1000 tons,	MG
QMSAC	Copper mine production capacity, South Africa and Namibia, in 1000 tons,	PD
QMTW	Copper mine production, total world, in 1000 tons,	MG
QMUS	Copper mine production, United States, in 1000 tons,	MG
QMUSC	Copper mine production capacity, United States, in 1000 tons,	PD
QMWW	Copper mine production, market economies, in 1000 tons,	MG
QMWWC	Copper mine production capacity, market economies, in 1000 tons,	PD
QMZI	Copper mine production, Zaire, in 1000 tons,	MG
QMZIC	Copper mine production capacity, Zaire, in 1000 tons,	PD
QMZM	Copper mine production, Zambia, in 1000 tons,	MG
QMZMC	Copper mine production capacity, Zambia, in 1000 tons,	PD
QRUS	Refined copper production, United States, in 1000 tons,	MG
QSGE	Production of refined copper from scrap, Federal Republic of Germany, in 1000 tons,	MG
QSJA	Production of refined copper from scrap, Japan, in 1000 tons,	MG

QSOM	Production of refined copper from scrap, other market economies, in 1000 tons,	MG
QSUS	Production of refined copper from scrap, United States, in 1000 tons,	MG
QSWW	Production of refined copper from scrap, market economies, in 1000 tons,	MG
R	Five year sample variance of PCULME/(REXUK*PCIF),	TR
STGE	Refined copper stocks, Federal Republic of Germany, in 1000 tons,	WBMS
STJA	Refined copper stocks, Japan, in 1000 tons,	WBMS
STLME	Refined copper stocks, London Metal Exchange, in 1000 tons,	WBMS
STOM	Refined copper stocks, other market economies, in 1000 tons,	WBMS
STUS	Refined copper stocks, United States, in 1000 tons,	WBMS
STWW	Refined copper stocks, market economies, in 1000 tons.	WBMS

Exogenous Variables

Variable	Description	Source
BSRW	Balancing surplus, not re-ported as stocks, outside the United States,	TR
BSUS	Balancing surplus, not re-ported as stocks, United States,	TR
COSTCA	Unit cost of mining index, Canada, 1975 = 1.00,	TR

COSTCH	Unit cost of mining index, Chile, 1975 = 1.00,	TR
COSTPE	Unit cost of mining index, Peru, 1975 = 1.00,	TR
COSTPH	Unit cost of mining index, Philippines, 1975 = 1.00,	TR
COSTSA	Unit cost of mining index, South Africa and Namibia, 1975 = 1.00,	TR
COSTUS	Unit cost of mining index, United States, 1975 = 1.00,	TR
COSTZI	Unit cost of mining index, Zaire, 1975 = 1.00,	TR
COSTZM	Unit cost of mining index, Zambia, 1975 = 1.00,	TR
DCH	Dummy variable for Chile before the nationalization of the Gran Minera: 1 in 1954-1969, 0 otherwise,	
D..	Dummy variable: 1 in 19.., 0 otherwise, (e.g.: D67 is 1 in 1967 and 0 otherwise),	
DEX	Dummy variable for U.S. copper export and price controls: 1 in 1965-1970 and 1973-1975, 0 otherwise,	
DPE	Dummy variable for Peru beginning with the expropriation of the Cerro Corp. properties by Centromin, the the State mining agency: 1 for 1974-1980, 0 otherwise,	
DSPEC	Dummy variable for the excessive speculation during the commodity boom 1972/73 and against the Pound sterling in 1978/1979: 1 in 1972/73 and 1979, 0 otherwise,	
DST	Dummy variable for strike induced changes in copper stocks: -.5 in 1954, 1968, 1971, -1 in 1967,	
DSTCA	Dummy variable for major strikes in the Canadian copper industry: .5 in 1966,1969,1978, 1 in 1979, 0 otherwise,	

DSTUS	Dummy variable for major strikes in the U.S. copper industry: .5 in 1959, 1968, 1975, 1979/80, 1 in 1967,	
DZI	Dummy variable for production cutbacks during the Katanga war and the dis- orders after it: 1 in 1963-1965, 0 otherwise,	
EUVI	Index of export unit values, indus- trialized countries, 1975=1.00,	IMF
EUVO	Index of export unit values, oil ex- porting countries, 1975=1.00,	IMF
IIPEW	Index of industrial production, centrally planned economies, 1975 = 100,	UN
IIPGE	Index of industrial production, Federal Republic of Germany, 1975 = 100,	IMF
IIPIT	Index of industrial production, Italy, 1975 = 100,	IMF
IIPJA	Index of industrial production, Japan, 1975 = 100,	IMF
IIPUK	Index of industrial production, United Kingdom, 1975 = 100,	IMF
IIPUS	Index of industrial production, United States, 1975 = 100,	IMF
IIPWW	Index of industrial production, market economies, 1975 = 100,	IMF
INTCA	Bank rate, end of period, Canada,	IMF
INTEU	Eurodollar rate in London (LIBOR),	IMF
INTGE	Discount rate, end of period, Federal Republic of Germany,	IMF
INTUK	Bank rate, United Kingdom, deflated with U.K. price index of industrial production,	IMF
PALLME	Average annual price of aluminum, 99.5 % ingot, London, lb. per ton,	MG
PALUS	Average annual price of aluminum, 99.5 % ingots, New York, U.S. $ per ton,	MG

PAU	Average annual price of gold, $ per fine ounce,	IMF
PCIF	International price index, unit values of manufactures (SITC 5-8), 1975 = 100,	WB
PWICA	Unit cost index of energy (fuel and electricity) and labor (wages and salaries), 1975 = 1.00,	CBS
PWICA1	Industry selling prices, Canada, 1975 = 1.00,	IMF
PWICH	Wholesale price index, industrial products, Chile, 1975=1.00,	UN
PWICH1	Wholesale price index, imported goods, Chile, 1975 = 1.00,	UN
PWIGE	Price index of industrial output, Federal Republic of Germany, 1975 = 1.00,	IMF
PWIIT	Wholesale price index, general, Italy, 1975 = 1.00,	IMF
PWIJA	Wholesale price index, general, Japan, 1975 = 1.00,	IMF
PWIPE	Index of consumer prices, Peru, 1975 = 1.00,	IMF
PWIPH	Wholesale price index, general, Philippines, 1975 = 1.00,	IMF
PWISA	Price index, home and import goods, South Africa, 1975 = 1.00,	IMF
PWIUK	Price index of industrial output, United Kingdom, 1975 = 1.00,	IMF
PWIUS	Wholesale price index, general, United States, 1975 = 1.00,	IMF
PWIUSE	Wholesale price index, electric power, 1975 = 1.00,	BLS1
PWIUSL	Average hourly earnings, copper production workers, United States, 1975 = 1.00,	BLS2

PWIUSM	Wholesale price index, mining and machinery, 1975 = 1.00,	BLS1
PWIZI	(PCIF*REXZI)/0.5	
PWIZM	Wholesale index general (ex-cluding copper), 1975 = 1.00,	CSO
REXCA	Exchange rate, Canadian Dollars per U.S. Dollar,	IMF
REXCH	Exchange rate, Chilean Pesos per U.S. Dollar,	IMF
REXGE	Exchange rate, Deutsche Mark per U.S. Dollar,	IMF
REXIT	Exchange rate, Italian Lire per U.S. Dollar,	IMF
REXJA	Exchange rate, Japanese Yen per U.S. Dollar,	IMF
REXPE	Exchange rate, Peruvian Soles per U.S. Dollar,	IMF
REXPH	Exchange rate, Philippine Pesos per U.S. Dollar,	IMF
REXSA	Exchange rate, Rand per U.S. Dollar,	IMF
REXUK	Exchange rate, Pound sterling per U.S. Dollar,	IMF
REXZI	Exchange rate, Zaires per U.S. Dollar,	IMF
REXZM	Exchange rate, Zambian Kwacha per U.S. Dollar,	IMF
STGUS	Refined copper stocks, strategic stockpile, United States,	WBMS
T	Time trend, 1963 = 10, increasing (decreasing) by 1 each succeeding (preceding) year,	
UCCA	User cost of capital index, mining, Canada, 1975 = 100,	TR
UCCH	User cost of capital index, mining, Chile, 1975 = 100,	TR
UCOM	User cost of capital index, mining, other market economies, 1975 = 100,	TR
UCPE	User cost of capital index, mining, Peru, 1975 = 100,	TR

UCPH	User cost of capital index, mining, Philippines, 1975 = 100,	TR
UCSA	User cost of capital index, mining, South Africa and Namibia, 1975 = 100,	TR
UCUS	User cost of capital index, mining, United States, 1975 = 100,	TR
UCZI	User cost of capital index, mining, Zaire, 1975 = 100,	TR
UCZM	User cost of capital index, mining, Zambia, 1975 = 100.	TR

Appendix III: Estimated Equations

Mine Production

Canada

$$\log(QMCA) = -.394 + .894 \log(QMCAC)$$
$$ (-1.68) \quad\quad (31.6)$$

$$+ .225 \log(PCUCA/COSTCA) - .175 \ DSTCA,$$
$$ (3.72) \quad\quad\quad\quad\quad\quad\quad\quad (-4.79)$$

$$\bar{R}^2 = .983, \quad\quad\quad\quad\quad DW = 1.29.$$

Chile

$$\log(QMCH) = -1.70 + 1.145 \log(QMCHC)$$
$$ (-3.62) \quad\quad (23.8)$$

$$+ .0682 \log(PCUCH*REXCH/COSTCH),$$
$$ (2.54)$$

$$\bar{R}^2 = .961, \quad\quad\quad\quad\quad DW = 1.58.$$

Peru

$$\log(QMPE) = -2.86 + 1.15 \log(QMPEC)$$
$$ (-3.17) \quad\quad (27.1)$$

$$+ .167 \log((PCULME*REXPE)/(REXUK*COSTPE)),$$
$$ (2.40)$$

$$\bar{R}^2 = .970, \quad\quad\quad\quad\quad DW = 1.78.$$

Philippines

$$\log(QMPH) = -.464 + .953 \log(QMPHC)$$
$$ (-1.67) \quad\quad (71.2)$$

$$+ .0657 \log((PCULME*REXPH)/(REXUK*COSTPH)),$$
$$ (2.37)$$

$$\bar{R}^2 = .995, \quad\quad\quad\quad\quad DW = 1.69.$$

The figures in parantheses are the asymptotic t-ratios. $\bar{R}^2$ denotes the adjusted coefficient of determination and DW means Durbin-Watson coefficient.

South Africa and Namibia

$$log(QMSA) = -.298 + .983\ log(QMSAC)$$
$$(-.845)\quad (36.2)$$

$$+ .0506\ log((PCULME*REXSA)/(REXUK*COSTSA)),$$
$$(1.13)$$

$$\bar{R}^2 = .981, \qquad\qquad DW = 1.59.$$

United States

$$log(QMUS) = -.693 + .873\ log(QMUSC)$$
$$(-1.38)\quad (22.4)$$

$$+ .202\ log(PCUUS/COSTUS) - .438\ DSTUS,$$
$$(3.90) \qquad\qquad (-14.2)$$

$$\bar{R}^2 = .96, \qquad\qquad DW = 1.90.$$

Zaire

$$log(QMZI) = .320 + .868\ log(QMZIC) - .149\ DZI$$
$$(1.39)\quad (33.4) \qquad\qquad (-6.97)$$

$$+ .0562\ log((PCULME*REXZI)/(REXUK*COSTZI)),$$
$$(3.12)$$

$$\bar{R}^2 = .981, \qquad\qquad DW = 2.49.$$

Zambia

$$log(QMZM) = -.234 + .908\ log(QMZMC)$$
$$(-.502)\quad (15.3)$$

$$+ .103\ log((PCULME*REXZM)/(REXUK*COSTZM)),$$
$$(3.44)$$

$$\bar{R}^2 = .904, \qquad\qquad DW = 1.80.$$

Other Market Economies

$$\log(\text{QMOM}) = \underset{(3.29)}{2.36} + \underset{(6.21)}{.620} \log(\text{QMOMC})$$

$$+ \underset{(1.27)}{.0490} \log(\text{PCULME}/(\text{REXUK}*\text{PCIF})),$$

Estimate of first-order autocorrelation coefficient: RHO = .869,
(8.76)

$$\bar{R}^2 = .607, \qquad\qquad \text{DW} = 1.47.$$

Centrally planned economies

$$\log(\text{QMEW}) = \underset{(22.2)}{4.48} + \underset{(2.78)}{.681} \log(\text{T}) + \underset{(.817)}{.167} \log(\text{IIPEW}),$$

Estimate of first-order autocorrelation coefficient: RHO = .635,
(4.11)

$$\bar{R}^2 = .967, \qquad\qquad \text{DW} = 1.98.$$

Mine Production Capacity

Canada

$$\text{QMCAC} = \underset{(0.10)}{2.38} + \underset{(24.3)}{.903} \text{QMCAC}_{-1}$$

$$+ \underset{(3.36)}{.178} \, (\text{PCUCA}*\text{QMCA}/\text{UCCA})$$

$$+ \underset{(.197)}{.0633} \, \text{R},$$

$$\bar{R}^2 = .980, \qquad\qquad \text{DW} = 1.73.$$

Chile

$$\text{QMCHC} = \underset{(1.61)}{155} + \underset{(9.38)}{.851} \text{QMCHC}_{-1}$$

$$+ \underset{(.494)}{.00137} \, (\text{PCUCH}*\text{QMCH}/\text{UCCH})$$

$$- \underset{(-1.00)}{.457} \, \text{R} \quad - \underset{(-1.70)}{53.3} \, \text{DCH},$$

$$\bar{R}^2 = .951, \qquad\qquad \text{DW} = 2.48.$$

Peru

$$QMPEC = \underset{(1.43)}{56.9} + \underset{(5.69)}{.724\ QMPEC_{-1}}$$

$$+ \underset{(.532)}{.000202}\ (PCULME*QMPE*REXPE/(REXUK*UCPE))$$

$$- \underset{(-1.09)}{.533\ R} + \underset{(2.42)}{60.8\ DPE,}$$

$$\bar{R}^2 = .873, \qquad DW = 2.57.$$

Philippines

$$QMPHC = \underset{(-.539)}{-3.99} + \underset{(37.9)}{1.00\ QMPHC_{-1}}$$

$$+ \underset{(1.69)}{.000508}\ ((PCULME*QMPH*REXPH/(REXUK*UCPH))$$

$$+ \underset{(.809)}{.110\ R,}$$

$$\bar{R}^2 = .987, \qquad DW = 1.53.$$

South Africa and Namibia

$$QMSAC = \underset{(-.261)}{-1.82} + \underset{(28.7)}{1.00\ QMSAC_{-1}}$$

$$+ \underset{(.945)}{.00239}\ (PCULME*QMSA*REXSA/(REXUK*UCSA))$$

$$+ \underset{(.244)}{.0269\ R} + \underset{(5.27)}{68.8\ D66,}$$

$$\bar{R}^2 = .978, \qquad DW = 1.90.$$

United States

$$QMUSC = \underset{(.843)}{31.6} + \underset{(43.1)}{.921\ QMUSC_{-1}}$$

$$+ \underset{(3.76)}{.00527}\ (PCUUS*QMUS/UCUS)$$

$$- \underset{(-.0173)}{.0055\ R} + \underset{(4.02)}{123\ D67} + \underset{(5.28)}{109\ (D75-D76),}$$

$$\bar{R}^2 = .990, \qquad DW = 2.36.$$

Zaire

$$QMZIC = \begin{array}{l} -5.72 \\ (-.252) \end{array} + \begin{array}{l} .995 \ QMZIC_{-1} \\ (23.8) \end{array}$$

$$+ \begin{array}{l} .00525 \quad (PCULME*QMZI*REXZI/(REXUK*UCZI)) \\ (1.89) \end{array}$$

$$- \begin{array}{l} .0108 \ R \\ (-.048) \end{array} - \begin{array}{l} 45.5 \ D78, \\ (-1.98) \end{array}$$

$$\bar{R}^2 = .964, \qquad\qquad DW = 1.24.$$

Zambia

$$QMZMC = \begin{array}{l} 80.7 \\ (2.57) \end{array} + \begin{array}{l} .896 \ QMZMC_{-1} \\ (23.1) \end{array}$$

$$+ \begin{array}{l} .000705 \quad (PCULME*QMZM*REXZM/(REXUK*UCZM)) \\ (.694) \end{array}$$

$$- \begin{array}{l} .258 \ R, \\ (-1.02) \end{array}$$

$$\bar{R}^2 = .971, \qquad\qquad DW = 1.80.$$

Other Market Economies

$$QMOMC = \begin{array}{l} -32.0 \\ (-.523) \end{array} + \begin{array}{l} 1.07 \ QMOMC_{-1} \\ (22.2) \end{array}$$

$$+ \begin{array}{l} .00108 \quad (PCULME*QMOM)/(REXUK*UCOM) \\ (.318) \end{array}$$

$$+ \begin{array}{l} .367 \ R, \\ (.497) \end{array}$$

$$\bar{R}^2 = .957, \qquad\qquad DW = 1.49.$$

Total Consumption

Federal Republic of Germany

$$\log(QCGE) = \underset{(7.35)}{2.71} - \underset{(-2.75)}{.0814} \log((PCULME_{-1}*REXGE_{-1})/(REXUK_{-1}*PWIGE_{-1}))$$

$$+ \underset{(3.67)}{.278} \log((PALLME_{-1}*REXGE_{-1})/(REXUK_{-1}*PWIGE_{-1}))$$

$$+ \underset{(19.8)}{.748} \log(IIPGE),$$

$$DW = 1.80.$$

Italy

$$\log(QCIT) = \underset{(1.57)}{1.50} - \underset{(-2.26)}{.076} \log((PCULME_{-1}*REXIT_{-1})/(REXUK_{-1}*PWIIT_{-1}))$$

$$+ \underset{(1.45)}{.140} \log((PALLME_{-1}*REXIT_{-1})/(REXUK_{-1}*PWIIT_{-1}))$$

$$+ \underset{(34.9)}{.892} \log(IIPIT),$$

$$DW = 1.70.$$

Japan

$$\log(QCJA) = \underset{(2.19)}{1.77} - \underset{(-2.49)}{.0972} \log((PCULME_{-1}*REXJA_{-1})/(REXUK_{-1}*PWIJA_{-1}))$$

$$+ \underset{(2.94)}{.305} \log((PALLME_{-1}*REXJA_{-1})/(REXUK_{-1}*PWIJA_{-1}))$$

$$+ \underset{(48.0)}{.826} \log(IIPJA),$$

$$DW = 1.94.$$

United Kingdom

$$\log(QCUK) = \underset{(1.85)}{1.08} - \underset{(-3.16)}{.0954} \log(PCULME_{-1}/PWIUK_{-1})$$

$$+ \underset{(9.51)}{1.37} \log(IIPUK),$$

$$- \underset{(-10.8)}{.0357} T,$$

$$DW = 1.17.$$

United States

$$\log(QCUS) = \underset{(7.84)}{5.58} - \underset{(-.308)}{.0346} \log(PCUUS_{-1}/PWIUS_{-1})$$

$$+ \underset{(.583)}{.129} \log(PALUS_{-1}/PWIUS_{-1})$$

$$+ \underset{(8.02)}{.461} \log(IIPUS) + \underset{(2.46)}{.203} D66,$$

$$DW = 1.45.$$

Other Market Economies

$$\log(QCOM) = \underset{(13.1)}{3.54} - \underset{(-2.92)}{.0851} \log(PCULME_{-1}/(REXUK_{-1}*PCIF_{-1}))$$

$$+ \underset{(1.67)}{.129} \log(PALLME_{-1}/(REXUK_{-1}*PCIF_{-1}))$$

$$+ \underset{(30.1)}{.912} \log(IIPWW),$$

$$DW = 1.29.$$

Refined Consumption

Federal Republic of Germany

$$QCRGE = \underset{(-.652)}{-21.0} + \underset{(19.3)}{.826} QCGE,$$

$$DW = 1.79.$$

Italy

$$QCRIT = \underset{(4.88)}{30.2} + \underset{(36.0)}{.565} QCIT,$$

$$DW = 1.10.$$

Japan

$$QCRJA = \begin{array}{c} -98.8 \\ (-10.2) \end{array} + \begin{array}{c} .808 \ QCJA, \\ (89.0) \end{array}$$

$$DW = .941.$$

United Kingdom

$$QCRUK = \begin{array}{c} 4.04 \\ (.143) \end{array} + \begin{array}{c} .779 \ QCUK, \\ (18.6) \end{array}$$

$$DW = 1.42.$$

United States

$$QCRUS = \begin{array}{c} - \ 248 \\ (-6.40) \end{array} + \begin{array}{c} .779 \ QCUS, \\ (51.4) \end{array}$$

$$DW = 1.52.$$

Other Market Economies

$$QCROM = \begin{array}{c} - \ 109 \\ (-3.70) \end{array} + \begin{array}{c} .822 \ QCOM, \\ (57.7) \end{array}$$

$$DW = 1.47.$$

Centrally Planned Economies

$$\log(QCREW) = \begin{array}{c} 4.59 \\ (34.4) \end{array} - \begin{array}{c} .0155 \ \log(PCULME_{-1}/(REXUK_{-1}*PCIF_{-1})) \\ (-.643) \end{array}$$

$$+ \begin{array}{c} .661 \ \log(IIPEW), \\ (25.1) \end{array}$$

Estimate of first-order autocorrelation coefficient: RHO = .628,
(4.03)

$$\bar{R}^2 = .963, \qquad\qquad DW = 2.04.$$

Demand for Storage

Federal Republic of Germany

$$STGE = \underset{(.238)}{1.96} + \underset{(1.78)}{.277} STGE_{-1} - \underset{(-1.00)}{2.01} INTGE$$

$$+ \underset{(1.74)}{.0325} (PCULMEF-PCULME)*REXGE/REXUK$$

$$+ \underset{(3.43)}{.695} IIPGE + \underset{(3.74)}{43.9} D77,$$

$$\bar{R}^2 = .888, \qquad\qquad DW = 2.00.$$

Japan

$$STJA = \underset{(-.0014)}{-.0216} + \underset{(3.80)}{.586} STJA_{-1}$$

$$+ \underset{(1.64)}{.0012} (PCULMEF-PCULME)*REXJA/REXUK$$

$$+ \underset{(2.05)}{.621} IIPJA + \underset{(1.44)}{59.0} D77,$$

$$\bar{R}^2 = .862, \qquad\qquad DW = 1.57.$$

London Metal Exchange

$$STLME = \underset{(-.0241)}{-.678} + \underset{(5.18)}{.353} STLME_{-1} - \underset{(-2.75)}{5.35} INTUK$$

$$+ \underset{(2.31)}{1.38} (PCULMEF-PCULME)$$

$$+ \underset{(1.88)}{.728} IIPWW + \underset{(7.90)}{278} (D75+D76+D77),$$

$$\bar{R}^2 = .969, \qquad\qquad DW = 2.54.$$

United States

$$STUS = \underset{(5.99)}{273} + \underset{(2.37)}{.232} STUS_{-1} - \underset{(-2.80)}{17.3} INTEU$$

$$+ \underset{(1.87)}{.533} (PCULMEF-PCULME)/REXUK - \underset{(-2.80)}{.139} STGUS$$

$$+ \underset{(2.46)}{1.22} EUVI - \underset{(-3.78)}{87.4} DSPEC + \underset{(7.73)}{117.6} DST,$$

$$\bar{R}^2 = .939, \qquad\qquad DW = 2.53.$$

Other Market Economies

STOM =   -9.39   + .734 STOM$_{-1}$
          (-.575)   (8.26)

            + .936 (PCULMEF-PCULME)/REXUK
               (.528)   -

            + .556 IIPWW   + 216 D75,
               (1.76)   (8.54)

            $\bar{R}^2$ = .961,   DW = 2.59.

Other Equations

Secondary Production

Federal Republic of Germany

QSGE =   30.4   + .219 QSGE$_{-1}$
           (1.42)   (1.24)

            + .439 ((PCULME*REXGE)/(REXUK*PWIGE))
               (2.25)

            + .804 IIPGE,
               (3.95)

            $\bar{R}^2$ = .682,   DW = 2.09.

Japan

QSJA =   -5.32   + .210 QSJA$_{-1}$
          (-.608)   (1.68)

            + .00365 ((PCULME*REXGE)/(REXUK*PWIGE))
               (2.95)

            + .515 QCJA,
               (6.12)

            $\bar{R}^2$ = .884,   DW = 1.65.

United States

$$QSUS = \begin{array}{c} -24.8 \\ (-.727) \end{array} + \begin{array}{c} .383 \ QSUS_{-1} \\ (4.23) \end{array}$$

$$+ \begin{array}{c} 8.30 \ PSUS/PWIUS \\ (3.26) \end{array} - \begin{array}{c} 505 \ STUS/QCUS \\ (-3.65) \end{array}$$

$$+ \begin{array}{c} 2.32 \ IIPUS \\ (6.81) \end{array} - \begin{array}{c} 41.1 \ D67, \\ (-1.37) \end{array}$$

$$\bar{R}^2 = .927, \qquad\qquad DW = 2.24.$$

Other Market Economies

$$QSOM = \begin{array}{c} 12.0 \\ (.523) \end{array} + \begin{array}{c} .623 \ QSOM_{-1} \\ (4.13) \end{array}$$

$$+ \begin{array}{c} 1.21 \ (PCULME/(REXUK*PCIF)) \\ (1.60) \end{array}$$

$$+ \begin{array}{c} 1.04 \ IIPWW \\ (2.25) \end{array}$$

$$\bar{R}^2 = .910, \qquad\qquad DW = 1.53.$$

Prices

Canada

$$PCUCA = \begin{array}{c} -10.1 \\ (-4.67) \end{array} + \begin{array}{c} .0171 \ (PCULME*REXCA/REXUK) \\ (6.74) \end{array}$$

$$+ \begin{array}{c} 54.3 \ COSTCA \\ (12.0) \end{array} + \begin{array}{c} 5.29 \ D70 \\ (1.32) \end{array} + \begin{array}{c} 6.94 \ D74, \\ (1.64) \end{array}$$

$$\bar{R}^2 = .975, \qquad\qquad DW = 1.26.$$

Chile

$$PCUCH = \begin{array}{c} -7.12 \\ (-.243) \end{array} + \begin{array}{c} 1.00 \ PCULME/REXUK \\ (43.5) \end{array} - \begin{array}{c} 226 \ (D64+2*D65+D66+D67), \\ (-10.3) \end{array}$$

$$\bar{R}^2 = .987, \qquad\qquad DW = 2.60.$$

London Metal Exchange: Spot Price

$$PCULME = -275 + .0871 \; QCWW - .0125 \; STWW_{-1}$$
$$(-3.14) \quad (4.12) \quad\quad (-.258)$$

$$- 13.0 \; T + 9.96 \; EUVI*REXUK + 148 \; DEX - 130 \; D78,$$
$$(-2.33) \quad (7.50) \quad\quad (6.04) \quad (-2.26)$$

$$\bar{R}^2 = .962, \quad\quad\quad DW = 1.53.$$

London Metal Exchange: Futures Price

$$PCULMEF = -92.7 + .913 \; PCULME$$
$$(-6.22) \quad (54.7)$$

$$+ 304 \; REXUK + .0756 \; PAU,$$
$$(6.47) \quad\quad (3.54)$$

$$\bar{R}^2 = .998, \quad\quad\quad DW = 1.94.$$

United States: Producer Price

$$PCUUS = 61.8 + .283 \; PCULME/REXUK$$
$$(1.10) \quad (3.98)$$

$$+ 952 \; COSTUS + 211 \; D70 + 212 \; D74,$$
$$(10.4) \quad\quad (2.04) \quad (1.86)$$

$$\bar{R}^2 = .952, \quad\quad\quad DW = 1.13.$$

United States: Scrap Price

$$PSUS = 85.3 + .585 \; PCULMEF/REXUK,$$
$$(2.33) \quad (20.2)$$

$$\bar{R}^2 = .942, \quad\quad\quad DW = 1.67.$$

East-West Trade

$$QIMWW = 8.76 + .556 \; QIMWW_{-1} + .320 \; (QMEW - QCREW)$$
$$(.175) \quad (3.46) \quad\quad (1.58)$$

$$- 2.60 \; PCULME/(REXUK*PCIF) + 1.57 \; IIPWW,$$
$$(-2.83) \quad\quad\quad (1.94)$$

$$\bar{R}^2 = .542, \quad\quad\quad DW = 2.19.$$

Refined Production: United States

$$QRUS = 1025 + .432\ QMUS - 1192\ STUS/QCUS - 319\ DSTUS,$$
$$(16.7)\quad (8.72)\quad\quad (-6.14)\quad\quad\quad (-8.56)$$

$$\bar{R}^2 = .906,\qquad\qquad DW = 1.57.$$

Identities

$QCRTW = QCREW + QCRWW,$

$QCRWW = QCRGE + QCRIT + QCRJA + QCROM + QCRUK + QCRUS,$

$QCWW = QCGE + QCIT + QCJA + QCOM + QCUK + QCUS,$

$QIMRUS = -BSUS + QCRUS - QMUS - QSUS + STGUS - STGUS_{-1} + STUS - STUS_{-1},$

$QIMUS = BSRW + QIMWW + QCUS - QCWW - QMUS + QMWW - QSUS + QSWW + STUS$
$\quad - STUS_{-1} - STWW + STWW_{-1},$

$QMTW = QMEW + QMWW,$

$QMWW = QMCA + QMCH + QMOM + QMPE + QMPH + QMSA + QMUS + QMZI + QMZM,$

$QMWWC = QMCAC + QMCHC + QMOMC + QMPEC + QMPHC + QMSAC + QMUSC + QMZIC$
$\quad + QMZMC,$

$QSWW = QSGE + QSJA + QSUS + QSOM,$

$STWW = STGE + STJA + STLME + STUS + STOM.$

Bibliography

Adams, F.G. (1973a)
The impact of copper production from the ocean floor. Study prepared for UNCTAD. Philadelphia: Economics Research Unit, University of Pennsylvania.

Adams, F.G. (1973b)
The integration of world primary commodity markets into Project LINK: The example of copper. Philadelphia: University of Pennsylvania. (Paper prepared for the annual meeting of Project LINK, Stockholm, September 3-8, 1973.)

Adams, F.G. (1975)
"Applied econometric modeling of non-ferrous metal markets. The case of ocean floor nodules." In W.A. Vogely (ed.): Mineral materials modeling. A state-of-the art review. Washington: Resources for the Future, 133-149.

Adams, F.G. (1977)
Commodity models in the LINK system. An empirical appraisal of commodity price impacts. Philadelphia: Department of Economics, Discussion Paper # 372, University of Pennsylvania.

Adams, F.G. (1978a)
"Primary commodity markets in a world model system." In F.G. Adams and S.A. Klein (eds.): Stabilizing World Commodity Markets. Lexington, Mass.: D.C. Heath and Company, 83-104.

Adams, F.G. (1978b)
"Impact of alternative commodity market stabilization policies on producing countries." American Statistical Association: Proceedings of the Business and Economic Statistics Section, 194-200.

Adams, F.G. (1978c)
Impact of manganese nodule production from the ocean floor: Long-term econometric estimates. Geneva: UNCTAD, TD/B/721/Add.1.

Adams, F.G. (1979a)
"Must high commodity prices depress the world economy: An application of a world model system." Journal of Policy Modeling, 1: 201-215.

Adams, F.G. (1979b)
"Integrating commodity models into LINK." In J.A. Sawyer (ed.): Modelling the international transmission mechanism. Applications and extensions of the Project LINK system. Amsterdam: North-Holland Publishing Company, 273-279.

Adams, F.G. (1980)
"The Law of the Sea Treaty and the regulation of nodule exploitation." Journal of Policy Modeling, 2: 19-33.

Adams, F.G., Behrman, J.R. (1981)
"The linkage effects of raw material processing in economic development: A survey of modeling and other approaches." Journal of Policy Modeling, 3: 375-397.

Adams, F.G., Behrman, J.R. (1982)
Commodity exports and economic development. Lexington, Mass.: D.C. Heath and Company.

Adams, R.G. (1979)
"The world economy to 1980." Journal of Metals, 5: 1979, 5-6.

Adams, R.G. (1981)
The Chase Econometric zinc forecasting model: A pragmatic approach. Discussion paper
presented to the National Academy of Sciences, Nonfuel Demand Modeling Workshop,
Airlie House, Warrenton, Virginia, June 2nd, 1981.

Alexandrov, E.A. (1980)
Mineral and energy resources of the USSR: A selected bibliography of sources in English. Falls Church, Va.: American Geological Institute.

Allingham, M. (1977)
"A model of the market for copper." In Association d'Econométrie Appliquée (ed.): 4.
Colloque International d'Econométrie Appliquée, Paris.

Allingham, M., Gilbert, C.L. (1977)
"Econometric modeling of the mineral sector with references to commodity agreements." In R.V. Ramani (ed.): Applications of computer methods in the mineral
industry. Proceedings of the fourteenth Symposium, October 4-8, 1976. New York,
1131-1143.

Anders, G., Gramm, W.P., Maurice, S.C., Smithson, C.W. (1980)
The economics of mineral extraction. New York: Praeger Publishing Company.

Anderson, R.C., Spiegelman, R.D. (1976)
Impact of the Federal Tax Code on resource recovery. Washington: Environmental Protection Agency.

Arrow, K.J., Chang, S.L.S. (1980)
"Optimal pricing, use and exploration of uncertain natural resource stocks." In P.T.
Liu (ed.): Dynamic optimization and mathematical economics. New York: Plenum Press,
105-116.

Arthur D. Little, Inc. (1976)
Economic impact of environmental regulations on the United States copper industry.
Draft report submitted to the Environmental Protection Agency.

Arthur D. Little, Inc. (1978)
Economic impact of environmental regulations on the United States copper industry.
Report submitted to the Environmental Protection Agency. Two Volumes.

Baak, T. (1978)
"Das schwierige Geschäft der Prognose - dargestellt am Beispiel des Paley-Reports."
Metall, 32: 831-833.

Bailey, G.L. (1966)
"Modern copper processing." In Norddeutsche Affinerie (ed.): Copper in nature, technics, arts and economics. Hamburg: Norddeutsche Affinerie A.G., 79-82.

Ballmer, R.W. (1960)
Copper market fluctuations: An industrial dynamics study. Unpublished M.S. Theses,
Massachussetts Institute of Technology.

Banks, F.E. (1969)
An economic and econometric analysis of the world copper market. Draft report prepared for UNCTAD, December 1969.

Banks, F.E. (1974)
The world copper market. An economic analysis. Cambridge, Mass.: Ballinger Publishing Company.

Banks, F.E. (1976)
"The price of copper." Intereconomics, 9: 248-251.

Banks, F.E. (1977)
"Natural resource availability: Some economic aspects." Resources Policy, 3: 2-12.

Barnett, H.J., Morse, C. (1963)
Scarcity and growth. The economics of natural resource availability. Baltimore: Johns Hopkins Press.

Barsotti, A.F., Rosenkranz, R.D. (1983)
"Estimated costs for the recovery of copper from demonstrated resources in market economy countries." Natural Resources Forum, 7(2): 99-113.

Beckerman, W. (1972)
"Economists, Scientists, and Environmental Catastrophe." Oxford Economic Papers, 24: 327-344.

Behrman, J.R. (1968)
Econometric models of mineral commodity markets: Uses and limitation. Proceedings of the Council of Economics, American Institute of Mining, Metallurgical and Petroleum Engineers, Inc., Annual Meeting, February 25-29, 1968, New York.

Behrman, J.R. (1970)
Forecasting properties and prototype simulations of a model of the copper market. Report prepared for the General Services Administration, Philadelphia: Economics Research Unit, University of Pennsylvania.

Behrman, J.R. (1975a)
Mini models for eleven international commodity markets. Report prepared for UNCTAD, Philadelphia: Economics Research Unit, University of Pennsylvania.

Behrman, J.R. (1975b)
Predictions of international prices for 28 commodities. Report prepared for UNCTAD, Philadelphia: Economics Research Unit, University of Pennsylvania.

Behrman, J.R. (1977)
International commodity agreements: An evaluation of the UNCTAD Integrated Commodity Programme. Washington: Overseas Development Council.

Behrman, J.R. (1978a)
Development, the international economic order, and commodity agreements. Reading, Mass.: Addison-Wesley Publishing Company.

Behrman, J.R. (1978b)
"International commodity agreements: An evaluation of the UNCTAD Integrated Programme for Commodities." In F.G. Adams and S.A. Klein (eds.): Stabilizing World Commodity Markets. Lexington, Mass.: D.C. Heath and Company, 295-321.

Behrman, J.R. (1979)
"Commodity agreements." In W.R. Cline (ed.): Proposals for a new international economic order: An economic analysis of effects on rich and poor countries. New York: Praeger Publishing Company, 61-153.

Behrman, J.R., Tinakorn, P. (1977)
The simulated potential gains of pooling international buffer stock financing across the UNCTAD core commodities. Report prepared for the Office of Raw Materials and Ocean Policy, U.S. Treasury, Philadelphia: University of Pennsylvania.

Behrman, J.R., Tinakorn, P. (1978)
"Evaluating integrated schemes for commodity market stabilization." In F.G. Adams and J.R. Behrman (eds.): Econometric modeling of world commodity policy. Lexington, Mass.: D.C. Heath and Company, 147-185.

Behrman, J.R., Tinakorn, P. (1979)
"Indexation of international commodity prices through international buffer stock operations." Journal of Policy Modeling, 1: 113-134.

Behrman, J.R., Tinakorn, P. (1980)
"The UNCTAD Integrated Programme: Earnings stabilization through buffer stocks for Latin America's commodities." In W.C. Labys, M.I. Nadiri and J.N. del Arco (eds.): Commodity markets and Latin American development: A modeling approach. Cambridge, Mass.: Ballinger Publishing Company, 245-274.

Bender, R.F. (1976)
"Metall-Rohstoffvorräte aus theoretischer und wirtschaftlicher Sicht." Beihefte der Konjunkturpolitik, 23: 9-50.

Bennett, H.J., Moore, L., Welborn, L.E., Toland, J.E. (1973)
An economic appraisal of the supply of copper from primary domestic sources. Information Circular 8598, Washington: U.S. Bureau of Mines.

Bennett, H.J., Thompson, J.G., Kingston, G.A. (1977)
"A systematic approach to the appraisal of national mineral supply." In R.V. Ramani (ed.): Applications of computer methods in the mineral industry. Proceedings of the fourteenth Symposium, October 4-8, 1976. New York, 45-55.

Bergsten, C.F. (1974)
"The new era in world commodity markets." Challenge, September/ October 1974, 34-42.

Bergsten, C.F. (1978)
American multinationals and American interests. Washington: The Brookings Institution.

Berndt, E.R., Morrison, C.J., Watkins, G.C. (1980)
Dynamic models of energy demand: An assessment and comparison. Resources Paper No. 49, University of British Columbia.

Bhaskar, K.N., Gilbert, C.L., Perlman, R.A. (1978)
"Stabilization of the international copper market." Resources Policy, 4: 13-24.

Billerbeck, K. (1975)
On negotiating a new order of the world copper market. Schriften des Deutschen Instituts für Entwicklungspolitik, No. 33, Berlin: German Development Institute.

Biswas, A.K., Davenport, W.G. (1976)
Extractive metallurgy of copper. Oxford, New York: Pergamon Press.

Bonczar, E.S., Tilton, J.E. (1975)
An economic analysis of the determinants of metal recycling in the United States: A case study of secondary copper. Open File Report 79-75, Washington: U.S. Bureau of Mines.

Bostock, M., Harvey, C. (1972)
"The takeover." In M. Bostock and C. Harvey (eds.): Economic independence and Zambian copper: A case study in foreign investment. New York: Praeger Publishing Company.

Boulding, K.E. (1966)
"The economics of the coming spaceship earth." In H. Jarrett (ed.): Environmental quality in a growing economy. Baltimore: Johns Hopkins Press.

Bourderie, J. (1974)
"Un métal rouge à réflets d'or. Les pays producteurs peuvent-ils réaliser pour le cuivre la même opération que l'O.P.E.C. pour le pétrole?" L'Economiste du Tiers Monde, 2: 11-16.

Bowen, R., Gunatilaka, A. (1977)
Copper, its geology and economics. London: Applied Science Publishers.

Bradley, C.E. (1948)
A statistical analysis of the demand for electrolytic copper with applications of simultaneous estimation of structural equations. Unpublished Ph.D. Thesis, University of Illinois.

Brobst, D.A. (1979)
"Fundamental concepts for the analysis of resource availability." In V.K. Smith (ed.): Scarcity and growth reconsidered. Baltimore: Johns Hopkins Press, 106-142.

Brown, G.M., Field, B.C. (1978)
"Implications of alternative measures of natural resource scarcity." Journal of Political Economy, 86: 229-243.

Brown, G.M., Field, B.C. (1979a)
"The adequacy of measures for signalling the scarcity of resources." In V.K. Smith (ed.): Scarcity and growth reconsidered. Baltimore: Johns Hopkins Press, 218-248.

Brown, G.M., Field, B.C. (1979b)
Possibilities for natural resource substitution in the U.S. economy. Washington: National Science Foundation.

Brundenius, K. (1972)
"The anatomy of imperialism: The case of the multinational mining corporations in Peru." Journal of Peace Research, 3: 189-207.

Budinski, K. (1979)
Engineering materials: Properties and selection. Reston, Va.: Reston Publishing Company.

Burgin, L.B. (1976)
Time required in developing selected Arizona copper mines. Information Circular 8702, Washington: U.S. Bureau of Mines.

Burley, S.P. (1974)
Short term variations in copper prices. Research Memorandum No. 82, Wien: Institut für Höhere Studien und Wissenschaftliche Forschung.

Burrows, J.C. (1975)
"Econometric modeling of metal and mineral industries." In W.A. Vogely (ed.): Mineral materials modeling. A state-of-the art review. Washington: Resources for the Future, 121-132.

Burrows, J.C., Lonoff, M.J. (1977)
Charles River Associates models of metal markets. Paper presented at the Ford Foundation Conference on Stabilizing World Commodity Markets, held at Airlie, Va., March 17, 1977.

Cammarota, V.A., Mo, W.J., Klein, B.W. (1980)
Projections and forecasts of U.S. mineral demand by the U.S. Bureau of Mines. Las Vegas, Nevada: 109th AIME Annual Meeting, February 24-29.

Carlson, J.W. (1975)
"Minerals models and policy decisions." In W.A. Vogely (ed.): Mineral materials modeling. A state-of-the art review. Washington: Resources for the Future.

Castelli, C. (1982)
Demand for copper in developed countries: 1955-1979. Washington: The World Bank, mimeo.

Chapman, P.F. (1974)
"Energy conservation and recycling of copper and aluminum." Metals and Materials, 8: 311-319.

Charles River Associates (1970)
Economic analysis of the copper industry. Washington: U.S. Department of Commerce.

Charles River Associates (1971)
The effects of pollution control on the non-ferrous metals industries. Copper. Part III. The economic impact of pollution abatement on the industry. Washington: U.S. Department of Commerce.

Charles River Associates (1973)
Forecasts and analysis of the copper market. Boston, Mass.: Charles River Associates, and Philadelphia, Pa.: Wharton Econometric Forecasting Associates.

Charles River Associates (1976)
Policy implications of producer country supply restrictions: The world copper market. Report submitted to the National Bureau of Standards, Boston, Mass.: Charles River Associates.

Charles River Associates (1977)
The feasibility of copper price stabilization using a buffer stock and supply restrictions from 1953 to 1976. Geneva: UNCTAD, TD/B/IPC/COPPER/AC/L.42.

Charles River Associates (1978)
Lead, copper and zinc price forecasts to 1985. Report submitted to the Environmental Protection Agency, Boston, Mass.: Charles River Associates.

Charles River Associates (1979)
Start-up of new mine, mill-concentrator, and processing plants for copper, lead, zinc and nickel: Survey and analysis. Boston, Mass.: Charles River Associates.

Chu, K.-Y. (1978)
"Short-run forecasting of commodity prices: An application of autoregressive moving average models." International Monetary Fund Staff Papers, 25: 90-111.

CIPEC (1974)
Modification de la Convention de CIPEC. Paris: Conseil Intergouvernemental des Pays Exportateurs de Cuivre, CM/37/74, July 24th, 1974.

Cissarz, A. et al. (1972)
Untersuchungen über Angebot und Nachfrage mineralischer Rohstoffe. 2. Kupfer. Hannover: Bundesanstalt für Bodenforschung, and Berlin: Deutsches Institut für Wirtschaftsforschung.

Citibank (1978)
"Copper prices and the OPEC that wasn't." Citibank Monthly Economic Letter, October 1978, 11-15.

Clarfield, K.W. et al. (1975)
Eight mineral cartels: The new challenge to industrialized nations. New York: McGraw Hill.

CODELCO (1982)
CODELCO-Chile. A profile. Santiago de Chile: Corporacion Nacional del Cobre de Chile.

Coen, R.M. (1968)
"Effects of tax policy on investment in manufacturing." American Economic Review, Papers and Proceedings, 58: 200-211.

Cohen, C. (1952)
Le cuivre et le nickel. Paris: Presses Universitaires de France.

Coleman, F.C. (1975)
The Northern Rhodesia copperbelt 1899-1962. Technical development up to the end of the Central African Federation. Manchester: Manchester University Press.

Commodities Research Unit, Ltd. (1978)
"A look at stock statistics." Copper Studies, 6: 1-8.

Commodities Research Unit, Ltd. (1981)
"The CRU long-term copper model." Copper Studies, 8: 1-7.

Commodity Exchange, Inc. (1977)
Statistical yearbook. New York: Commodity Exchange, Inc..

Commodity Exchange, Inc. (1982)
Copper futures. New York: Commodity Exchange, Inc..

Cooper, R.N. (1975)
"Resource needs revisited." Brookings Papers on Economic Activity, 1 (1975): 238-245.

Copper Development Association
Copper supply and consumption. Annual.

Cordero, H.G., Tarring, L.H. (1960)
Babylon to Birmingham. London: Quin Press.

Cox, D.P. et al. (1973)
"Copper." In D.A. Brobst and W.P. Pratt (eds.): United States Mineral Resources. Geological Survey Paper No. 820, Washington: Government Printing Office.

Crowson, P. (1979)
"The geography and political economy of metal supplies." Resources Policy, 5: 158-169.

Cuddy, J.D.A. (1976)
Consideration of issues relating to the establishment and operation of a common fund, financial requirements. Geneva: UNCTAD, TD/B/IPC/CF/L.2.

Cunningham, S. (1981)
The copper industry in Zambia. New York: Praeger Publishing Company.

Dammert, A. (1977)
A world copper model for project design. Unpublished Ph.D. Dissertation, University of Texas at Austin.

Dammert, A. (1980)
"Planning investments in the copper sector in Latin America." In W.C. Labys, M.I. Nadiri and J.N. del Arco (eds.): Commodity markets and Latin American development: A modeling approach. Cambridge, Mass.: Ballinger Publishing Company, 65-83.

Dammert, A., Kendrick, D. (1976)
A world copper model. Department of Economics Discussion Paper # 8076, Austin: University of Texas.

Dasgupta, P., Heal, G.M. (1979)
Economic theory and exhaustible resources. Cambridge, England: Cambridge University Press.

Davidoff, R.L. (1980)
Supply analysis model (SAM): A minerals availability system methodology. Information Circular 8820, Washington: U.S. Bureau of Mines.

Davis, D. (1975)
"Secondary copper recovery." Resources Policy, 1: 246-252.

Dayton, S. (1979)
"CIPEC's Big Four: Chile, Zambia, Zaire, Peru." Engineering and Mining Journal, 1980 (November 1979): 66-76, 79-88, 91-98, 101-106, 114-118, 121-130, 133-142, 146-152, 155-156, 159-178, 181-182, 188-196, 199-206.

Deffaa, W. (1982)
"Die Berücksichtigung monopolistischer und oliopolistischer Strukturen in der statistischen Konzantrationsmessung." Allgemeines Statistisches Archiv, 66: 323-340.

Demler, F.R., Tilton, J. (1980)
"Modeling international trade flows in mineral markets, with implications for Latin America's trade policies." In W.C. Labys, M.I. Nadiri and J.N. del Arco (eds.): Commodity markets and Latin American development: A modeling approach. Cambridge, Mass.: Ballinger Publishing Company, 85-120.

Devarajan, S., Fisher, A.C. (1981)
"Hotelling's 'Economics of exhaustible resources': Fifty years later." Journal of Economic Literature, 19: 65-73.

Ehrlich, P.R., Ehrlich, A.H., Holdren, J.P. (1977)
Ecoscience, population, resources, environment. San Francisco: W.H. Freeman and Company.

Elliott, W.Y. et al. (1937)
International control in the non-ferrous metals. New York: Macmillan Company.

Ertek, T. (1967)
World demand for copper, 1948-63. An econometric study. Unpublished Ph.D. Dissertation, University of Wisconsin.

Faber, M.L.O., Potter, J.G. (1971)
Towards economic interdependence: Papers on the nationalisation of the copper industry in Zambia. Cambridge, England: Cambridge University Press.

Fairchild Publications
Metal statistics. The purchasing guide of the metal industries. New York: Fairchild Publications, Annual.

Faundez, J. (1978)
"A decision without a strategy: Excess profits in the nationalization of copper in Chile." In J. Faundez and S. Picciotto (eds.): Nationalization of multinationals in peripheral economies. New York: Holmes and Meier Publishers.

Felgran, S.D. (1982)
Producer prices versus market prices in the world copper industry. Unpublished Ph.D. Dissertation, Yale University.

Fink, W.H. (1948)
Copper cartels and prices: 1915-1947. Unpublished Ph.D. Dissertation, University of California at Berkeley.

Fisher, A.C. (1979)
"Measures of natural resource scarcity." In V.K. Smith (ed.): Scarcity and growth reconsidered. Baltimore: Johns Hopkins University Press, 249-275.

Fisher, F.M., Cootner, P.H. Baily, M. (1972)
"An econometric model of the world copper industry." Bell Journal of Economics, 3: 568-609.

Fletcher, J.S. (1981)
A computer simulation model of the economic stockpiling of copper employing an automatic mechanism for the adjustment of prices. Unpublished Ph.D. Dissertation, Pennsylvania State University.

Flinn, R.A., Trojan, P.K. (1975)
Engineering materials and their application. Boston, Mass.: Houghton Mifflin Company.

Foders, F., Kim, C. (1982)
"Perspektiven des Manganmarktes am Vorabend des Tiefseebergbaus. Eine Analyse alternativer Szenarien." Weltwirtschaft, Heft 1: 75-94.

Foders, F., Kim, C. (1983)
A simulation model for the world manganese market. Kieler Arbeitspapiere, Working Paper No. 170, Institut für Weltwirtschaft an der Universität Kiel.

Foley, P. (1979)
A supply curve for the domestic primary copper industry. Unpublished Ph.D. Dissertation, Massachusetts Institute of Technology.

Foley, P., Clark, J. (1981)
"U.S. copper supply. An economic / engineering analysis of cost-supply relationships." Resources Policy, 7: 171-187.

Fry, J., Harvey, C. (1974)
"Copper and Zambia." In S. Pearson and J. Cownie (eds.): Commodity exports and African economic development. Lexington, Mass.: D.C. Heath and Company.

Gaffey, M.J., McCord, T.B. (1977)
"Mining outer space." Technology Review, 79: 50-59.

Ghosh, S., Gilbert, C.L., Hughes Hallett, A.J. (1981a)
A model of the world copper industry. Rotterdam: Erasmus University, Institute for Economic Research, Discussion Paper 8107.

Ghosh, S., Gilbert, C.L., Hughes Hallett, A.J. (1981b)
Optimal control and choice of functional form: An application to a model of the world copper industry. Rotterdam: Erasmus University, Institute for Economic Research, Discussion Paper 8109.

Ghosh, S., Gilbert, C.L., Hughes Hallett, A.J. (1981c)
A quarterly model of the world copper industry: Some further results. Rotterdam: Erasmus University, Institute for Economic Research, Discussion Paper 8114.

Ghosh, S., Gilbert, C.L., Hughes Hallett, A.J. (1982a)
Rational expectations and commodity market modeling. An application to a model of the world copper market. Mimeo, April 1982.

Ghosh, S., Gilbert, C.L., Hughes Hallett, A.J. (1982b)
Commodity market stabilization. A comparison of simple and optimal intervention strategies in a model of the world copper market. Mimeo, June 1982.

Ghosh, S., Gilbert, C.L., Hughes Hallett, A.J. (1982c)
"Optimal stabilization of the copper market." Resources Policy, 8: 201-214.

Gilbert, C.L. (1982)
Some problems in testing the efficient market hypothesis of foreign reserve and commodity market data. Oxford: Institute of Economics and Statistics, University of Oxford, October 1982.

Girvan, N. (1970)
Copper in Chile: A study in conflict between corporate and national economy. Kingston: University of the West Indies.

Gluschke, W., Shaw, J., Varon, B. (1979)
Copper. The next fifteen years. Dordrecht: D. Reidel Publishing Company.

Goeller, H.E. (1979)
"The age of substitutability: A scientific appraisal of natural resource adequacy." In V.K. Smith (ed.): Scarcity and growth reconsidered. Baltimore: Johns Hopkins University Press.

Goeller, H.E., Weinberg, A.V. (1976)
"The age of substitutability." Science, 191: 683-689.

Goeller, H.E., Weinberg, A.V. (1978)
"The age of substitutability." American Economic Review, 68: 1-11.

Gordon, R.L. (1967)
"A reinterpretation of the pure theory of exhaustion." Journal of Political Economy, 75: 274-286.

Gordon, R.L., Lambo, W.A., Schenck, G.H.K. (1972a)
Effective systems of scrap utilization: Copper, aluminum, nickel. Report submitted to the U.S. Bureau of Mines, University Park: Pennsylvania State University.

Gordon, R.L., Lambo, W.A., Schenck, G.H.K. (1972b)
The collection of non-ferrous scrap. A literature review of the copper and aluminum sector. Report submitted to the U.S. Bureau of Mines, University Park: Pennsylvania State University.

Goss, B.A. (1981)
"The forward pricing function of the London Metal Exchange." Applied Economics, 13: 133-150.

Govett, G.J.S., Govett, M.H. (1974)
"The concepts and measurement of mineral reserves and resources." Resources Policy, 1: 46-55.

Govett, G.J.S., Govett, M.H. (eds.) (1976)
World mineral supplies. Assessment and perspective. Amsterdam, Oxford: Elsevier Scientific Publishing Company.

Govett, G.J.S., Govett, M.H. (1977)
"Scarcity of basic minerals and fuels: Assessments and implications." In D.W. Pearce and I. Walter (eds.): Resource conservation. Social and economic dimensions of recycling. New York: New York University Press, 33-63.

Grace, R.P. (1978)
"Metals recycling. A comparative national analysis." Resources Policy, 4: 249-256.

Grubb, T.J. (1981)
"Metal inventories, speculation, and stability in the U.S. copper industry." Materials and Society, 5: 267-288.

Grund, U. (1981)
"Schrotteinsatz in der NE-Metallindustrie." Metall, 35: 1275-1276.

Gueronik, S.R. (1970)
The Intergovernmental Council of Copper Exporting Countries (CIPEC). Paris: CIPEC.

Gueronik, S.R. (1974)
"CIPEC and the semis." In Copper. A Metal Bulletin Special Issue. London: Metal Bulletin Limited, 41-47.

Gueronik, S.R. (1975)
"Sieben Jahre CIPEC (1968-1975)." Metall, 29: 1151-1153.

Gupta, P. (1981)
An econometric model of the world cobalt industry. Kiel Working Paper No. 129, Kiel: Institut für Weltwirtschaft an der Universität Kiel.

Gupta, S., Mayer, T. (1981)
"A test of the efficiency of futures markets in commodities." Weltwirtschaftliches Archiv, 117: 661-671.

Habenicht, H. (1977)
"Processing raw materials." Intereconomics, 9/10: 230-232.

Habig, G. (1983)
Möglichkeiten und Grenzen einer Kontrolle internationaler Rohstoffmärkte durch Entwicklungsländer. Das Beispiel des Kupfer- und Aluminiummarktes. Hamburg: Verlag Weltarchiv.

Hall, R.E., Jorgenson, D.W. (1967)
"Tax policy and investment behavior." American Economic Review, 57: 391-414.

Hansen, L.P. (1978)
Econometric modeling strategies for exhaustible resource markets with applications to non-ferrous metals. Unpublished Ph.D. Dissertation, University of Minnesota.

Hartman, R.S. (1977a)
An oligopolistic pricing model of the U.S. copper industry. Unpublished Ph.D. Dissertation, Massachusetts Institute of Technology.

Hartman, R.S. (1977b)
A critical survey of three copper industry models and their policy uses. Energy Laboratory Working Paper # MIT-EL 77-028WD, Massachusetts Institute of Technology.

Hartman, R.S. (1980)
"Some evidence on differential inventory behavior in competitive and noncompetitive settings." Quarterly Review of Economics and Business, 20: 11-27.

Hartman, R.S., Bozdogan, K., Nadkarni, R. (1979)
"The economic impacts of environmental regulations on the U.S. copper industry." Bell Journal of Economics, 10: 589-618.

Heal, G.M., Barrow, M. (1980)
"Metal price movements and interest rates." Review of Economic Studies, 47: 161-182.

Hendry, D.F. (1980)
"Econometrics: Alchemy or science?" Economica, 47: 387-406.

Herfindahl, O.C. (1959)
Copper costs and prices: 1870-1957. Baltimore: Johns Hopkins University Press.

Herfindahl, O.C. (1967)
"Depletion and economic theory." In M. Gaffney (ed.): Extractive resources and taxation. Madison: University of Wisconsin Press, 63-90.

Hibbard, W.R., Soyster, A.L., Gates, R.S. (1980a)
A disaggregated engineering supply model of the U.S. copper industry operating in an aggregated world econometric supply / demand system. Washington: U.S. Bureau of Mines.

Hibbard, W.R., Soyster, A.L., Gates, R.S. (1980b)
"A disaggregated engineering supply model of the U.S. copper industry operating in an aggregated world econometric supply / demand system." Materials and Society, 4: 261-284.

Hicks, J.R. (1946)
Value and capital. Second edition, Oxford: Clarendon Press.

Hojman, D.E. (1980)
"International trade in copper: Static and dynamic aspects of the instability problem and the commodity agreement solution." In A. Sengupta (ed.): Commodities, finance and trade: Issues in North-South negotiations. London: Frances Pinter.

Hojman, D.E. (1982)
"Chilean mining. Perverse supply response to profit rate incentives." Resources Policy, 8: 75-77.

Hotelling, H. (1931)
"The economics of exhaustible resources." Journal of Political Economy, 392: 137-175.

Howe, C.W. (1979)
Natural resource economics. Issues, analysis and policy. New York: John Wiley and Sons.

Hu, S.D. (1978)
The copper commodity model and energy issues. Unpublished Ph.D. Dissertation, University of Pennsylvania.

Hu, S.D., Zandi, I. (1980)
"Copper econometric models." Engineering costs and production economics, 5: 53-70.

Hwa, E.C. (1979)
"Price determination in several international primary commodity markets: A structural analysis." International Monetary Fund Staff Papers, 26: 157-188.

Ilunkamba, I. (1980)
"Copper technology and dependence in Zaire: Towards the demystification of the new white magic." Natural Resources Forum, 4: 147-156.

Iskander, W.B. (1973)
The economic impact of copper in the Canadian Arctic. Unpublished Ph.D. Dissertation, Indiana University.

Jaksch, H.J. (1982)
"Oligopolistic supply on the world cocoa market." In W. Eichhorn (ed.): Economic theory and natural resources. Würzburg: Physica-Verlag, 187-203.

Jankovic, S. (1967)
Wirtschaftsgeologie der Erze. Wien, New York: Springer Verlag.

Johnson, M.H., Bell, F.W., Bennett, J.T. (1980)
"Natural resource scarcity: Empirical evidence and public policy." Journal of Environmental Economics and Management, 7: 256-271.

Johnson, M.H., Bennett, J.T. (1980)
"Increasing resource scarcity: Further evidence." Quarterly Review of Economics and Business, 20: 42-48.

Joralemon, I.B. (1973)
Copper. The encompassing story of mankind's first metal. Berkeley: Howell-North Books.

Judge, G.G. et al. (1980)
The theory and practice of econometrics. New York: John Wiley and Sons.

Kellman, M. (1978)
A copper market model. Philadelphia: Wharton Econometric Forecasting Associates.

Khanna, I. (1972)
"Forecasting the price of copper", The Business Economist, 4(1), 33-40.

Klass, M.W., Burrows, J.C., Beggs, S.D. et al. (1980)
International mineral cartels and embargoes. Policy implications for the United States. New York: Praeger Publishing Company.

Klein, L.R., Young, R.M. (1980)
An introduction to econometric forecasting and forecasting models. Lexington, Mass.: D.C. Heath and Company.

Klein, S. (1968)
Preliminary estimates of the world copper market. Philadelphia: Economics Service Unit, University of Pennsylvania.

Königliche Bergacademie zu Freiberg (ed.) (1845)
Kalender für den Sächsischen Berg= und Hütten= Mann auf das Jahr 1845. Dresden: Druck der Teubner'schen Officin.

Kovisars, L. (1975)
World trade flows in copper. Palo Alto: Stanford Research Institute.

Krasner, S.D. (1974)
"Oil is the exception." Foreign Policy, 14: 68-84.

Kraume, E. (1966)
"Copper ore deposits." In Norddeutsche Affinerie (ed.): Copper in nature, technics, arts and economics. Hamburg: Norddeutsche Affinerie A.G., 17-20.

Krauss, U. (1976)
Joint studies on supply of and demand for mineral raw materials carried out by the Federal Institute for Geosciences and Natural Resources, Hannover, and the German Institute for Economic Research, Berlin. September 1976, mimeo.

Kravis, I.B., Zoltan, K., Heston, A., Summers, R. (1975)
Phase I: A system of international comparisons of gross product and purchasing power. Baltimore: Johns Hopkins University Press.

Kravis, I.B., Heston, A., Summers, R. (1978)
Phase II: International comparisons of real product and purchasing power. Baltimore: Johns Hopkins University Press.

174

Kravis, I.B., Heston, A., Summers, R. (1982)
Phase III: World product and income: International comparisons of real GDP. Baltimore: Johns Hopkins University Press.

Kuijper, M.A.M. de (1983)
The unraveling of market regimes in theory and in application to copper, aluminum and oil. Unpublished Ph.D. Dissertation, Harvard University.

Kusik, C.L., Kenahan, C.B. (1978)
Energy use patterns for metal recycling. Information Circular 8781, Washington: U.S. Bureau of Mines.

Labys, W.C. (1973)
Dynamic commodity models: Specification, estimation and simulation. Lexington, Mass.: D.C. Heath and Company.

Labys, W.C. (1977)
Minerals commodity modeling: The state of the art. Washington: Council of Economics of the AIME, Proceedings of the Mineral Economics Symposium on Minerals Policies in Transition, 80-106.

Labys, W.C. (1978)
"Bibliography of commodity models." In F.G. Adams and J.R. Behrman (eds.): Econometric modeling of world commodity policy. Lexington, Mass.: D.C. Heath and Company, 187-223.

Labys, W.C. (1980a)
Market structure, bargaining power, and resource price formation. Lexington, Mass.: D.C. Heath and Company.

Labys, W.C. (1980b)
"A model of disequilibrium adjustments in the copper market." Materials and Society, 4: 153-164.

Labys, W.C. (1982)
"The role of state trading in mineral commodity markets." In M.M. Kostecki (ed.): State trading in international markets. New York: St. Martin's Press, 78-102.

Labys, W.C., Afrasiabi, A. (1983)
"Cyclical disequilibrium in the U.S. copper market." Applied Economics, 15: 437-449.

Labys, W.C., Granger, C.W.G. (1970)
Speculation, hedging and commodity price forecasts. Lexington, Mass.: D.C. Heath and Company.

Labys, W.C., Kaboudan, M.A. (1980)
Quarterly model of disequilibrium adjustment in the copper market. NSF Working paper, West Virginia University.

Labys, W.C., Rees, H.J.B., Elliott, C.M. (1971)
"Copper price behavior and the London Metal Exchange." Applied Economics, 3: 99-113.

Labys, W.C., Thomas, H.C. (1975)
"Speculation, hedging and commodity price behavior." Applied Economics, 7: 287-301.

Lasaga, M. (1979)
An econometric analysis of the impact of copper on the Chilean economy. Unpublished Ph.D. Dissertation, University of Pennsylvania.

Lasaga, M. (1981)
The copper industry in the Chilean economy. Lexington, Mass.: D.C. Heath and Company.

Law, A.D. (1975)
International commodity agreements: Setting, performance and prospects. Lexington, Mass.: D.C. Heath and Company.

Lee, S. (1980)
Buffer stock rules for world commodity markets: An application of optimal control theory. Unpublished Ph.D. Dissertation, Cornell University.

Lee, S., Blandford, D. (1980a)
"An analysis of international buffer stocks for cocoa and copper through dynamic optimization." Journal of Policy Modeling, 2: 371-388.

Lee, S., Blandford, D. (1980b)
Buffer stock price stabilization: An application of optimal control theory. Ithaca, New York: Agricultural Experimental Station, Cornell University.

Leland, H.Y. (1980)
"Alternative long-run goals and the theory of the firm: Why profit maximization may be a better assumption than you think." In P.T. Liu (ed.): Dynamic optimization and mathematical economics. New York, London: Plenum Press, 31-50.

Lenoble, J.-P. (1981)
"Polymetallic nodules resources and reserves in the North Pacific from the data collected by AFERNOD." Ocean Management, 7: 9-24.

Lindert, P.H., Kindleberger, C.P. (1982)
International economics. Seventh edition, Homewood, Ill.: Richard D. Irwin, Inc..

Lira, R. (1974)
The impact of an export commodity in a developing economy: The case of the Chilean copper 1956-1968. Unpublished Ph.D. Dissertation, University of Pennsylvania.

Lira, R. (1980)
"The impact of copper in the Chilean economy." In W.C. Labys, M.I. Nadiri and J.N. del Arco (eds.): Commodity markets and Latin American development: A modeling approach. Cambridge, Mass.: Ballinger Publishing Company, 185-210.

Lonoff, M. (1978)
Introduction to a model of long-run copper supply. Boston, Mass.: Charles River Associates.

Lonoff, M. (1980)
"Looking at copper supply: A long-run perspective." Materials and Society, 4: 165-176.

Lonoff, M., Reddy, B. (1979)
Price formation mechanisms in econometric models of metals markets: A comparative survey. Paper presented at the RFF/EPRI Conference on Natural Resource Prices, Boston, Mass.: Charles River Associates.

Mackenzie, B.W. (1979)
Canada's competitive position in copper and zinc markets. Centre for Resource Studies, Working Paper No. 16, Kingston, Ontario: Queen's University.

MacKinnon, J.G., Olewiler, N.D. (1980)
"Disequilibrium estimation of the demand for copper." Bell Journal of Economics, 11: 197-211.

Maddala, G.S. (1977)
Econometrics. New York: McGraw-Hill.

Mahalingasivam, R. (1969)
Market for Canadian refined copper. Unpublished Ph.D. Dissertation, University of Toronto.

Malenbaum, W. (1973)
Materials requirements in the United States and abroad in the year 2000. Philadelphia: University of Pennsylvania.

Malenbaum, W. (1977)
World demand for raw materials in 1985 and in 2000. Philadelphia: University of Pennsylvania.

Mardones, J.L., Marshall, I., Silva, E. (1980)
"Copper price stabilization: Welfare consequences and buffer stock costs." Natural Resources Forum, 4: 292-305.

Mariano, R.S. (1978)
"Commodity market modeling: Methodological issues and control theory applications." In F.G. Adams and J.R. Behrman (eds.): Econometric modeling of world commodity policy. Lexington, Mass.: D.C. Heath and Company.

Martner, G. (1979)
Producers - exporters associations of developing countries. An instrument for the establishment of a new international economic order. Geneva.

May, E.S. (1937)
"The copper industry in the United States." In W. Elliott et al. (eds.): International control in the non-ferrous metals. New York: Macmillan Company.

Massachusetts Institute of Technology (1978)
Effects of public regulation on the U.S. copper industry. Volume IV. Background information and analysis. Cambridge, Mass.: Center for Policy Alternatives, Massachusetts Institute of Technology.

McCarthy, J.L. (1963)
The American copper industry, 1918-1955. Unpublished Ph.D. Dissertation, Yale University.

McCarthy, J.L. (1964)
"The American copper industry, 1947-1955." Yale Economic Essays, 4: 65-132.

McGill, S. (1983)
"Project financing applied to the Ok Tedi mine. A government perspective." Natural Resources Forum, 7: 115-129.

McKern, R.B. (1981)
"The industrial economics of copper processing." Natural Resources Forum, 5: 227-248.

McKinnon, R.I. (1967)
"Futures markets, buffer stocks, and income stability for primary producers." Journal of Political Economy, 75: 844-861.

McNicol, D.L. (1973)
The two price system in the copper industry. Unpublished Ph.D. Dissertation, Massachusetts Institute of Tehnology.

McNicol, D.L. (1975)
"The two price system in the copper industry." Bell Journal of Economics, 6: 50-73.

Meadows, D. et al. (1972)
The limits to growth. A report for the Club of Rome's project on the predicament of mankind. New York: Universe Books.

Merbach, K. (1881)
"über die Anlagen zur Unschädlichmachung des Rauches auf den fiscalischen Hüttenwerken bei Freiberg." Jahrbuch für das Berg- und Hüttenwesen im Königreiche Sachsen auf das Jahr 1881. Freiberg: Craz & Gerlach, 42-49.

Metallgesellschaft A.G. (ed.)
Metal Statistics. Frankfurt am Main: Metallgesellschaft Aktiengesellschaft. Annual.

Mezger, D. (1977)
Konflikt und Allianz in der internationalen Rohstoffwirtschaft: das Beispiel Kupfer. Bonn, Bremen: Bremer Afrika Archiv und Progress dritte welt verlag.

Mezger, D. (1980)
Copper in the world economy. New York, London: Monthly Review Press.

Mikdashi, Z. (1974)
"Collusion could work." Foreign Policy, 14: 57-58.

Mikesell, R.F. (1974)
"More Third World cartels ahead?" Challenge, 17 (November/ December 1974): 24-31.

Mikesell, R.F. (1975)
Foreign investment in copper mining: Case studies of mines in Peru and Papua New Guinea. Baltimore: Johns Hopkins University Press.

Mikesell, R.F. (1979)
The world copper industry. Structure and economic analysis. Baltimore: Johns Hopkins University Press.

Mikesell, R.F. (1980)
"The structure of the world copper industry." In S. Sideri and S. Johns (eds.): Mining for development in the Third World. New York, Oxford, Toronto: Pergamon Press, 61-69.

Milliken, F. (1966)
"Modern copper mining." In Norddeutsche Affinerie (ed.): Copper in nature, technics, arts and economics. Hamburg: Norddeutsche Affinerie A.G., 55-62.

Mingst, K.A. (1976)
"Cooperation or illusion: An examination of the Intergovernmental Council of Copper Exporting Countries." International Organization, 30: 263-287.

Mingst, K.A., Stauffer, R.E. (1979)
"Intervention analysis of political disturbances, market shocks, and policy initiatives in international commodity markets." International Organization, 33: 105-118.

Mo, W.Y., Klein, B.W. (1980)
Bureau of Mines statistical projections methodology of U.S. mineral consumption by end-use: Aluminum as an example. Information Circular 8825, Washington: U.S. Bureau of Mines.

Montluc, M., Techeur, P. (1978)
"Le marché mondial du cuivre." Etudes et Expansion, 77: 581-602.

Moran, T.H. (1974)
Multinational corporations and the policies of dependence. Copper in Chile. Princeton: Princeton University Press.

Muhly, J.D. (1973)
"Copper and tin. The distribution of mineral resources and the nature of the metal
trade in the Bronze Age." In Connecticut Academy of Arts and Sciences (ed.): Transactions, New Haven, Conn., 155-535.

Müller-Ohlsen, L. (1981)
Die Weltmetallwirtschaft im industriellen Entwicklungsprozeß. Tübingen: Verlag
J.C.B. Moor.

Multinational Business (1978)
Investment crisis in mining industries: The case of copper. Multinational Business,
1: 19-30.

Murapa, R. (1976)
"Nationalization of the Zambian mining industry." Review of Black Political Economy,
7: 40-52.

Nahum, R. (1975)
"Der italienische Metallhandel und die Börse." Metall, 29: 520-522.

Nappi, C. (1979)
Commodity market controls. Lexington, Mass.: D.C. Heath and Company.

Navin, T.R. (1978)
Copper mining and management. Tucson, Arizona.

Netschert, B.C., Landsberg, H.H. (1978)
The future supply of the major metals: A reconnaissance survey. Westwood, Conn.:
Greenwood Press. (Reprint of the 1961 edition published by Resources for the Fu-
ture.)

Newbery, D.M.G., Stiglitz, J.E. (1981)
The theory of commodity price stabilization. A study in the economics of risk. London: Oxford University Press.

Newhouse, J.P., Sloan, F.A. (1966)
An econometric study of copper supply. Santa Monica, Ca.: Rand Corporation.

Nordhaus, W.D. (1973)
"World dynamics - Measurement without data." Economic Journal, 82: 1156-1183.

Nordhaus, W.D. (1974)
"Resources as a constraint to growth." American Economic Review, 64: 22-26.

Nziramasanga, M.T. (1973)
The copper export sector and the Zambian economy. Unpublished Ph.D. Dissertation,
Stanford University.

Obernolte, W. (1980)
"Überlegungen zu den Kupfergesprächen in der UNCTAD." Metall, 34: 67-69.

Obidegwu, C.F. (1980)
The impact of copper price movements and variability on the Zambian economy: A
dynamic analysis. Unpublished Ph.D. Dissertation, University of Pennsylvania.

Obidegwu, C.F., Nziramasanga, M. (1981a)
Copper and Zambia: An econometric analysis. Lexington, Mass.: D.C. Heath and Company.

Obidegwu, C.F., Nziramasanga, M.T. (1981b)
"Primary commodity price fluctuations and developing countries: An econometric model of copper and Zambia." Journal of Development Economics, 9: 89-119.

Ogawa, K. (1982a)
Expectations and price stabilization policy of primary commodities. Unpublished Ph.D. Dissertation, University of Pennsylvania.

Ogawa, K. (1982b)
A new approach to econometric modeling of primary commodities: A world copper model. Philadelphia: University of Pennsylvania, mimeo.

Oi, W.Y. (1961)
"The desirability of price instability under perfect competition." Econometrica, 29: 58-64.

Panayotou, T. (1979a)
"OPEC as a model for copper exporters: Potential gains and cartel behavior." The Developing Economies, 17: 203-219.

Panayotou, T. (1979b)
The copper cartel and Canada: Likelihood and implications of OPEC-type strategies. Resources Paper # 32, Vancouver: University of British Columbia.

Park, C.F. (1975)
Earthbound. San Francisco, Ca.: Freeman-Cooper.

Patiño, H.C. de (1965)
Las fluctuaciones del mercado del cobre. Soluciones alternativas. Santiago de Chile: Universitad de Chile.

Perlman, R.A. (1977)
Copper stockpile modeling. Paper presented at the Conference on Stabilizing World Commodity Markets at Airlie, Va., March 17-20, 1977.

Perlman, R.A. (1982)
"Kupferverbrauch in Entwicklungsländern." Metall, 36: 80-82.

Peterson, F.M., Fisher, A.C. (1977)
"The exploitation of extractive resources: A survey." Economic Journal, 87: 681-721.

Pethig, R. (1979)
"Die Knappheit natürlicher Ressourcen." Jahrbuch für Sozialwissenschaft, 30: 189-209.

Petrick, A., Bennett, H.J., Starch, K.E., Weisner, R.C. (1973)
The economics of byproduct metals. 1. Copper System. Information Circular 8569, Washington: U.S. Bureau of Mines.

Pickard, W.C., Krumm, R.J. (1977)
"Joint aluminum - copper forecasting and simulation model." Proceedings of the Council of Economics. New York: American Institute of Mining, Metallurgical, and Petroleum Engineers, Inc., 77-87.

Piesch, W. (1975)
Statistische Konzentrationsmaße. Formale Eigenschaften und verteilungstheoretische Zusammenhänge. Tübingen: Verlag J.C.B. Mohr.

Piesch, W. (1982)
"Einige statistische Aspekte des konzentrationsanalytischen Studienprogramms der Europäischen Gemeinschaften." In W. Piesch and W. Förster (eds.): Angewandte Statistik und Wirtschaftsforschung heute. Göttingen: Vandenhoek & Ruprecht.

Pindyck, R.S. (1978a)
"Gains to producers from the cartelization of exhaustible resources." Review of Economics and Statistics, 60: 238-251.

Pindyck, R.S. (1978b)
"The optimal exploration and production of nonrenewable resources." Journal of Political Economy, 86: 841-861.

Pobukadee, J. (1980)
An econometric analysis of the world copper market. Philadelphia: Wharton Econometric Forecasting Associates.

Puller, J.G. (1971)
"The 51 per cent nationalisation of the Zambian copper mines." In M.L.O. Faber and J.G. Potter (eds.): Towards economic independence. Papers on the nationalisation of the copper industry in Zambia. Cambridge, England: Cambridge University Press, 91-134.

Prain, R. (1980)
Copper. The anatomy of an industry. London: Mining Journal Books.

Preston, R.S. (1972)
The Wharton annual and industry forecasting model. Economics Research Unit, University of Pennsylvania.

Price Waterhouse
Corporate taxes in 80 countries. New York: Price Waterhouse.

Price Waterhouse Information Guide (1979)
Doing business in Peru. Lima: Price Waterhouse and Company.

Prior, K. (1966)
"Modern copper smelting and refining." In Norddeutsche Affinerie (ed.): Copper in nature, technics, arts and economics. Hamburg: Norddeutsche Affinerie A.G., 63-76.

Prokop, F.W. (1975)
The future economic significance of large copper and nickel deposits. Monograph Series on Mineral Deposits, Berlin: Gebrüder Borntrager.

Pugh-Roberts Associates (1976)
Materials system policy making. A case study of copper. Report to the Science and Technology Policy Office, Contract No. NSF - C977.

Radetzki, M. (1970)
International commodity market agreements. A study of the effects of post-war commodity agreements and compensatory finance. London: C. Hurst.

Radetzki, M. (1975)
"Metal mineral resource exhaustion and the threat to material progress: The case of copper." World Development, 3: 123-136.

Radetzki, M. (1977a)
"Long-term copper production options of the developing countries." Natural Resources Forum, 1: 145-155.

Radetzki, M. (1977b)
Will the long-run global supply of industrial metal minerals be adequate? Seminar Paper No. 85, Stockholm: Institute for International Economic Studies, University of Stockholm.

Radetzki, M. (1978)
"Market structure and bargaining power. A study of three international markets."

Stockholm: Institute for International Economic Studies, Reprint Series No. 95. Reprinted from Resources Policy, June 1978.

Radetzki, M. (1979)
"The rising costs of base materials. The case of copper." Mining Magazine, April 1979, 351-359.

Radetzki, M. (1980)
"Changing structures in the financing of the minerals industry in the LDCs." Development and Change, 11: 1-15.

Radetzki, M. (1981)
Has political risk scared mineral investments away from the deposits in developing countries? Seminar Paper No. 169, Stockholm: Institute for International Economic Studies, University of Stockholm.

Radetzki, M. (1983)
"Long-run price prospects for aluminium and copper."
Natural Resources Forum, 7: 23-36.

Radetzki, M., Svensson, L.E.O. (1979)
"Can scrap save us from depletion?" Natural Resources Forum, 3: 365-378.

Radetzki, M., Zorn, S.A. (1979)
Financing mining projects in developing countries. London: Mining Journal Books. (A United Nations Study.)

Radetzki, M., Zorn, S.A. (1980)
"Foreign finance for LDC mining projects." In S. Sideri and S.Johns (eds.): Mining for development in the Third World. New York: Pergamon Press, 177-197.

Rafati, M.R. (1982a)
Potentielle Auswirkungen des Tiefseebergbaus auf den Nickelmarkt. Eine Modellprognose. Die Weltwirtschaft, Heft 1, 95-109.

Rafati, M.R. (1982b)
An econometric model of the world nickel industry. Kieler Arbeitspapiere, Working Paper No. 160, Institut für Weltwirtschaft an der Universität Kiel.

Rafati, M.R. (1982c)
An econometric model of the world cobalt industry. Kieler Arbeitspapiere, Working Paper No. 163, Institut für Weltwirtschaft an der Universität Kiel.

Rampacek, C. (1977)
"The impact of R&D on the utilization of low-grade resources." Chemical Engineering Progress, 73: 57-68.

Resources for Freedom (1952)
A report to the President by the President's Materials Policy Commission. Volume II, Washington: Government Printing Office.

Rham, G. (1977)
La politique étrangère de la République de Zambie. Bern: P. Lang.

Richard, D. (1978)
"A dynamic model of the world copper industry." International Monetary Fund Staff Papers, 25: 779-833.

Ridge, J.D. (1976)
Annotated bibliographies of mineral deposits in Africa, Asia (exclusive of the USSR) and Australasia. New York, Oxford Toronto: Pergamon Press.

Ridker, R.G. (ed.) (1972)
Population, resources, and the environment. Washington: Government Printing Office.

Robbins, P. (1982)
Guide to non-ferrous metals and their markets. Third edition. New York: Nichols; London: Kogan Page.

Rohatgi, P.K., Weiss, C. (1977)
"Technology forecasting for commodity projections: A case study on the effect of substitution by aluminum on the future demand for copper." Technological Forecasting and Social Change, 22: 25-46.

Rosenkranz, R.D. (1976)
Energy consumption in domestic primary copper production. Information Circular 8693, Washington: U.S. Bureau of Mines.

Rosenkranz, R.D., Davidoff, R.L., Lemons, J.F. (1979)
Copper availability - domestic. A minerals availability appraisal. Information Circular 8809, Washington: U.S. Bureau of Mines.

Rumberger, M., Wettig, E. (1978)
Rohstoff Kupfer. Strukturen und Trends bei Angebot und Nachfrage. Bonn: Forschungs-institut der Friedrich-Ebert-Stiftung.

Salisbury, H.B., Duchene, L.J., Bilbrey, J.H. (1981)
Recovery of copper and associated precious metals from electronic scrap. RI 8561, Washington: U.S. Bureau of Mines.

Sawyer, J.A. (ed.) (1979)
Modelling the international transmission mechanism. Applications and extensions of the Project LINK system. Amsterdam: North-Holland.

Sawyer, J.A., Obidegwu, C. (1979)
"Incorporating commodity models into national models. A copper model." In Sawyer, J.A. (ed.): Modelling the international transmission mechanism. Applications and extensions of the Project LINK system. Amsterdam: North-Holland, 295-308.

Schlager, K.J. (1961)
A systems analysis of the copper and aluminum industries. An industrial dynamics study. Unpublished M.S. Thesis, Massachusetts Institute of Technology.

Schneider, H.K. (1980)
"Implikationen der Theorie erschöpfbarer natürlicher Ressourcen für wirtschaftspoli-tisches Handeln." In H. Siebert (ed.): Erschöpfbare Ressourcen. Schriften des Vereins für Socialpolitik, N.F. Bd. 108, Berlin: Duncker & Humblot, 815-844.

Schröder, A. (1966)
"Copper occurrences in nature." In Norddeutsche Affinerie (ed.): Copper in nature, technics, arts and economics. Hamburg: Norddeutsche Affinerie A.G., 12-16.

Schroeder, H.J. (1979)
Copper. Mineral Commodity Profile, Washington: U.S. Bureau of Mines.

Schroeder, H.J., Jolly, J.H. (1981)
Copper. Preprint from Bulletin 671, Washington: U.S. Bureau of Mines.

Seidman, A. (ed.) (1975)
Natural resources and national welfare. The case of copper. New York, Washington, London: Praeger Publishers.

Selten, R. (1973)
"A simple model of imperfect competition, where 4 are few and 6 are many." International Journal of Game Theory, 2: 141-201.

Sharpe, W.F. (1978)
Investments. Englewood Cliffs, N.J.: Prentice Hall.

Shaw, J.F. (1978)
"Investment in the copper industry: Needs and politics." Natural Resources Forum, 2:
101-118.

Sideri, S., Johns, S. (eds.) (1980)
Mining for development in the Third World. Multinational corporations, state enterprises and the international economy. New York: Pergamon Press.

Siebert, H. (1979)
"Indikatoren der Knappheit natürlicher Ressourcen." Wirtschaftsdienst, 1979/ VII:
409-416.

Siebert, H. (ed.) (1980)
Erschöpfbare Ressourcen. Schriften des Vereins für Socialpolitik, N.F., Bd. 108,
Berlin: Duncker & Humblot.

Siebert, H. (1981)
Economics of the environment. Lexington, Mass.: D.C. Heath and Company.

Siebert, H. (1983)
ökonomische Theorie natürlicher Ressourcen. Tübingen: Verlag J.C.B. Mohr.

Sies, W. (1981a)
"Welthandel mit Erzen und Metallen." Technische Mitteilungen, 74: 151-158.

Sies, W. (1981b)
"Die NE-Metalle im Jahr 1980 und im bisherigen Verlauf des Jahres 1981." In Metall-
gesellschaft A.G. (ed.) Metallstatistik 1970-1980, Frankfurt am Main:
Metallgesellschaft A.G., V-X.

Sies, W. (1981c)
"Führt der geringere DM - Außenwert zu höheren Rohstoffkosten?" Metall, 35: 783.

Silber, S. (1982)
"Acquisitions and mergers." Materials and Society, 6: 31-41.

Silverman, B.G. (1977)
Resoumetric modeling: Copper - A case study. Unpublished Ph.D. Dissertation, University of Pennsylvania.

Skelton, A. (1937)
"Copper." In W.Y. Elliott et al. (eds.): International control in the non-ferrous
metals. New York: Macmillan Company, 363-536.

Skinner, B.J. (1976)
"A second Iron Age ahead?" American Scientist, 64: 258-269.

Slade, M.E. (1979)
An econometric model of the domestic copper and aluminum industry: The effects of
higher energy prices and declining ore quality on metal substitution and recycling.
Unpublished Ph.D. Dissertation, George Washington University.

Slade, M.E. (1980a)
"The effects of higher energy prices and declining ore quality. Copper - aluminum
substitution and recycling in the USA." Resources Policy, 6: 223-239.

Slade, M.E. (1980b)
"Price changes and metals markets: Modeling short- and long-run copper - aluminum substitution." Materials and Society, 4: 397-411.

Slade, M.E. (1980c)
"An econometric model of the U.S. secondary copper industry: Recycling versus disposal." Journal of Environmental Economics and Management, 7: 123-141.

Smart, J.S. (1954)
"The physical properties of copper." In A. Butts (ed.): Copper, the metal, its alloys and compounds. New York: Reinhold Publishing Company.

Smith, G.W. (1975)
An econometric evaluation of international buffer stocks for copper. Philadelphia: Wharton Econometric Forecasting Associates, Draft.

Smith, M.J. (1966)
"Typical uses of copper and its alloys." In Norddeutsche Affinerie (ed.): Copper in nature, technics, arts and economics. Hamburg: Norddeutsche Affinerie A.G., 83-92.

Smith, V.K. (1981)
"The empirical relevance of Hotelling's model for natural resources." Resources and Energy, 3: 105-117.

Smithson, C.W. et al. (1979)
World mineral markets: An econometric and simulation analysis. Ontario Ministry of Natural Resources.

Smithson, C.W. et al. (1981a)
World mineral markets. Stage II. An econometric and simulation approach to world markets for copper, aluminum, nickel and zinc. Ontario Ministry of Natural Resources.

Smithson, C.W. et al. (1981b)
World mineral markets. Stage III. An econometric and simulation approach to world markets for copper, aluminum, nickel and zinc. Ontario Ministry of Natural Resources.

Solveen, K. (1977)
"Die internationalen Verhandlungen für Kupfer." Metall, 31: 78-79.

Sousa, L.J. (1981)
The U.S. copper industry. Problems, issues, and outlook. Washington: U.S. Bureau of Mines.

Soyster, A.L., Sherali, H.D. (1981)
"On the influence of market structure in modelling the US copper industry." Omega, 9: 381-388.

Sparrow, F.T., Soyster, A.L. (1980)
Process models of minerals industries. Paper presented at the 109th Annual Meeting of the American Institute of Mining, Metallurgical and Petroleum Engineers, Inc., held in Las Vegas, Nevada, February 24-29, 1980.

Spendlove, M. (1961)
Methods of producing secondary copper. Washington: U.S. Department of Commerce, Bureau of Standards.

Spendlove, M. (1969)
Opportunities in the production of secondary non-ferrous metals. New York: UNIDO.

Staloff, S.J. (1977)
A stock-flow-analysis of copper markets. Unpublished Ph.D. Dissertation, University of Oregon.

Stamper, J.W., Kurz, H.F. (1978)
Aluminum. Mineral Commodity Profiles MCP-14, Washington: U.S. Bureau of Mines.

Standard & Poor's Industry Surveys (1981)
Metals - Nonferrous. Basic analysis. Section 2, May 21, 1981.

Stevens, B.J. (1977)
Impact of the Federal Tax Code on resource recovery. A condensation. New York: Columbia University.

Stewardson, B.R. (1970)
"The nature of competition in the world market for refined copper." Economic Record, 46: 169-181.

Streissler, E. (1980)
"Die Knappheitsthese. Begründete Vermutungen oder vermutete Fakten?" In H. Siebert (ed.): Erschöpfbare Ressourcen. Schriften des Vereins für Socialpolitik, N.F. Bd. 108, Berlin: Duncker & Humblot, 9-36.

Streit, M.E. (1980)
"Zur Funktionsweise von Terminkontraktmärkten." Jahrbücher für Nationalökonomie und Statistik, 195: 533-549.

Strishkov, V.V. (1979)
"The copper industry of the USSR." Mining Magazine, March 1979: 242-253, and May 1979: 429-441.

Strongman, J.E., Killingsworth, W.R., Cummings, W.E. (1977)
"The dynamics of the international copper system." In R.V. Ramani (ed.): Application of computer methods in the mineral industry. Proceedings of the fourteenth Symposium, October 4-8, 1976, New York, 688-700.

Strongman, J.E., Killingsworth, W.R., Cummings, W.E. (1978)
Materials policy analysis. A case study of copper. Final Report to the Office of Science and Technology Policy, National Science Foundation, Contract # C-977, revised version, Cambridge, Mass.: Pugh-Roberts Associates.

Sutulov, A. (1967)
Copper production in Russia. Concepcion: University of Concepcion.

Sutulov, A. (1979)
"Chilean copper resources." Natural Resources Forum, 3: 211-215.

Sutolov, A. (1982)
World copper outlook 1983-2000. Santiago de Chile: Intermet Publications.

Synergy, Inc. (1975)
A forecasting system for critical imported materials. Final Report prepared for the U.S. Bureau of Mines, Washington: Synergy, Inc.

Synergy, Inc. (1977)
Joint aluminum / copper forecasting and simulation model. Report and appendices. Washington: U.S. Bureau of Mines.

Takeuchi, K. (1972)
"CIPEC and the copper export earnings of member countries." The Developing Economies, 10: 1-29.

Takeuchi, K. (1975)
"Copper: Market prospects for 1980 and 1985." In A. Seidman (ed.): Natural resources and national welfare. The case of copper. New York, Washington, London: Praeger Publishers, 55-59.

Tanzer, M. (1980)
The race for resources. Continuing struggles over minerals and fuels. New York, London: Monthly Review Press.

Tatsch, J.H. (1975)
Copper deposits: Origins, evolution and present characteristics. Sudbury, Mass.: Tatsch Associates.

Taylor, C.A. (1979)
"A quarterly domestic copper industry model." Review of Economics and Statistics, 61: 410-422.

Thiebach, G. (1976)
Copper: Current situation and outlook for 1976. World Commodity Paper No. 18, Washington: The World Bank.

Thiebach, G., Helterline, R. (1978)
Copper: Current situation and short-term outlook. Commodity Note No. 3, Washington: The World Bank.

Thomas, T.C. (1962)
Variations in the copper usage in the United States. Unpublished Ph.D. Dissertation, Massachusetts Institute of Technology.

Tilton, J.E. (1966)
The choice of trading partners: An analysis of international trade in aluminum, bauxite, copper, lead, manganese, tin, and zinc. Unpublished Ph.D. Dissertation, Yale University.

Tilton, J.E. (1977a)
The future of nonfuel minerals. Washington: The Brookings Institution.

Tilton, J.E. (1977b)
"The continuing debate over the exhaustion of nonfuel mineral resources." Natural Resources Forum, 1: 167-173.

Tilton, J.E. (1981)
"Cyclical instability: A growing threat to metal producers and consumers." Natural Resources Forum, 5: 5-14.

Tims, W., Singh, S. (1977)
A system of linked models for commodity market analysis. Commodity Note No. 8, Washington: The World Bank.

Tinakorn, P. (1978)
An evaluation of the UNCTAD Integrated Commodity Program. Unpublished Ph.D. Dissertation, University of Pennsylvania.

Tinsley, C.R. (1977)
"Computer applications of non-ferrous econometric models from the raw materials consumer perspective." In R.V. Ramani (ed.): Application of computer methods in the mineral industry. Proceedings of the fourteenth Symposium, October 4-8, 1976, New York, 701-713.

Tomimatsu, T.T. (1980)
The U.S. copper mining industry. A perspective of financial health. Information Circular 8836, Washington: U.S. Bureau of Mines.

Turnovsky, S.J. (1978)
"The distribution of welfare gains from price stabilization: A survey of some theoretical issues." In F.G. Adams and S. Klein (eds.): Stabilizing world commodity markets. Lexington, Mass.: D.C. Heath and Company.

Turnovsky, S.J. (1979)
"Futures markets, private storage, and price stabilization." Journal of Public Economics, 12: 301-327.

UNCTAD (1974)
An intergrated programme for commodities. TD/B/C.1/166, Geneva: UNCTAD.

UNCTAD (1977a)
Marketing and pricing methods for copper. TD/B/IPC/COPPER/AC/L.15, Geneva: UNCTAD.

UNCTAD (1977b)
Long-term trends in the demand and supply of copper. TD/B/IPC/COPPER/AC/L.23, Geneva: UNCTAD.

UNCTAD (1977c)
Substitution including the relative stability of supplies and prices of copper and competing metals. TD/B/IPC/COPPER/AC/L.25, Geneva: UNCTAD.

Underwood, J.M. (1976)
Optimization rules for producer groups in a stochastic market setting with application to the copper and tea markets. Unpublished Ph.D. Dissertation, University of Minnesota.

Underwood, J.M. (1977)
"Optimal rules for cartel managers with empirical applications to the copper and tea markets." Annals of Economic and Social Measurement, 6: 231-243.

United States Bureau of Mines
Minerals yearbook. Washington: U.S. Government Printing Office. Annual.

United States Bureau of Mines
Mineral Commodity Summaries. Washington: U.S. Government Printing Office.

United States Bureau of Mines
Mineral Facts and Problems. Washington: U.S. Government Printing Office.

United States Geological Survey (1980)
Principles of a resource / reserve classification for minerals. Circular 831, Washington: U.S. Geological Survey.

United Nations (1972)
Copper production in developing countries. New York: United Nations.

United Nations (1981)
Transnational corporations in the copper industry. New York: Centre on Transnational Corporations, United Nations.

Varian, H.R. (1978)
Microeconomic analysis. New York: Norton.

Varon, B., Gluschke, W. (1977)
"Mineral commodity projections as a tool for planning." In R.V. Ramani (ed.): Application of computer methods in the mineral industry. Proceedings of the fourteenth Symposium, October 4-8, 1976, New York, 1103-1120.

Vokes, F.M. (1976)
"The abundance and availability of mineral resources." In Govett, G.J.S., Govett, M.H. (eds.): World mineral supplies. Assessment and perspective. Amsterdam, Oxford: Elsevier Scientific Publishing Company, 65-97.

Von Broich-Oppert, K.O.F. (1926)
Der Kampf der Kontinente um den Kupfermarkt. Inaugural= Dissertation zur Erlangung der Doktorwürde einer Hohen Philosophischen Fakultät der Universität Leipzig.

Waelbroek, J.L. (ed.) (1976)
The models of Project LINK. Amsterdam: North-Holland.

Wagenhals, G. (1981a)
Private storage and public buffer stocks. Some preliminary thoughts. Working Paper, Philadelphia: Economics Research Unit, University of Pennsylvania.

Wagenhals, G. (1981b)
Buffer stocks for non-ferrous metals? Working Paper, Philadelphia: Economics Research Unit, University of Pennsylvania.

Wagenhals, G. (1982)
"Zur Theorie der Rohstoffpreisstabilisierung." Jahrbücher für Nationalökonomie und Statistik, 197/5: 263-269.

Wagenhals, G. (1983a)
The world copper market: Structure and econometric model. Final Report for the Deutsche Forschungsgemeinschaft, Heidelberg: Alfred Weber-Institut, University of Heidelberg, Unpublished Manuscript, February 1983.

Wagenhals, G. (1983b)
A bibliography of econometric non-ferrous metal market models. Heidelberg: Alfred Weber-Institut, University of Heidelberg, March 1983.

Wagenhals, G. (1983c)
"ökonometrische Modelle für NE-Metallmärkte." Metall, 37(5): 514-519.

Wagenhals, G. (1984a)
"Econometric models of minerals markets - Uses and limitations." Natural Resources Forum, 8(1): 77-86.

Wagenhals, G. (1984b)
The Impact of Copper Production from Manganese Nodules. A Simulation Study. Kieler Arbeitspapiere, Working Paper No. 192, Institut für Weltwirtschaft an der Universität Kiel.

Wagenhals, G. (1984c)
"New Developments in Commodity Market Modeling." In P. Thoft-Christensen (ed.): System Modelling and Optimization. Lecture Notes in Control and Information Sciences, Vol. 59, New York, Berlin, Heidelberg: Springer-Verlag, 62-70.

Wagenhals, G. (1984d)
Möglichkeiten und Grenzen internationaler Rohstoffkartelle - das Beispiel Kupfer. Paper prepared for the Conference of the Verein für Socialpolitik, Travemünde, September 15th-17th, 1984.

Wagenhals, G. (1984e)
"Erfolgsaussichten eines CIPEC-Kupferkartells." Forthcoming in Metall, 38(11), November 1984.

Wang, K.P. (1977)
Mineral resources and basic industries in the People's Republic of China. Boulder, Colorado: Westview Press.

Waugh, F.V. (1944)
"Does the consumer benefit from price instability?" Quarterly Journal of Economics, 58: 602-614.

Wenk, E. (1969)
"The physical resources of the ocean." Scientific American, 221 (September 1969): 166-176.

White, L. (1984)
"Copper from the Swedish Artic." Engineering and Mining Journal, February 1984, 29-33.

Whitney, J.D. (1854)
The metallic wealth of the United States compared with that of other countries. Philadelphia: Lippincott, Gramba and Company.

Whitney, J.W. (1976)
An analysis of copper production, processing, and trade patterns. Unpublished Ph.D. Dissertation, Pennsylvania State University.

Williams, D. (1983)
"Historical survey. Copper prices 1784-1982." Cycles, 34(3): 71-78.

Williams, D.H., Brodice, J.N., Poulter, D.M. (1973)
"Evaluation of production strategies in a group of copper mines by linear programming." In M.D.G. Salmon and F.H. Lancaster (eds.): 10th symposium on application of computer methods in the mineral industry. Johannesburg, 285-289.

Wilson, P.R.D. (1977)
Export instability and economic development: A survey. Warwick Economic Research Papers No. 107 and 111, Warwick: University of Warwick.

Wimpfen, S.P., Bennett, H.J. (1975)
"Copper resources appraisal." Resources Policy, 1: 126-141.

Wolff, F.F. (1975)
"The London Metal Exchange. Ruhender Pol in einer sich ständig verändernden Welt." Metall, 29: 199.

Wolff and Co. Ltd. (1980)
Wolff's guide to the London Metal Exchange. 2nd ed., Worcester Park, Surrey: Metal Bulletin Books Limited.

World Bank (1979)
Chile. An economy in transition. Washington: The World Bank.

World Bank (1981)
World Bank commodity models. Washington: The World Bank.

World Bank (1982)
1981 World Bank atlas. Gross national product, population and growth rates. Washington: The World Bank.

World Bureau of Metal Statistics (1977)
World copper statistics since 1950. Birmingham, England: World Bureau of Metal Statistics.

World Bureau of Metal Statistics
World Metal Statistics. Birmingham, England: World Bureau of Metal Statistics. Quarterly.

Zellner, A. (1962)
"An efficient method of estimating seemingly unrelated regressions and tests for aggregation bias." Journal of the American Statistical Association, 57: 348-368.

Zorn, S. (1978)
"Producers' associations and commodity markets: The case of CIPEC." In F.G. Adams and S. Klein (eds.): Stabilizing world commodity markets. Lexington, Mass.: D.C. Heath and Company, 215-234.

Zuckerman, E. (1979)
"The mining of the moon." Boston, 71: 20, 22-23.

Vol. 213: Aspiration Levels in Bargaining and Economic Decision Making. Proceedings, 1982. Edited by R. Tietz. VIII, 406 pages. 1983.

Vol. 214: M. Faber, H. Niemes und G. Stephan, Entropie, Umweltschutz und Rohstoffverbrauch. IX, 181 Seiten. 1983.

Vol. 215: Semi-Infinite Programming and Applications. Proceedings, 1981. Edited by A. V. Fiacco and K. O. Kortanek. XI, 322 pages. 1983.

Vol. 216: H. H. Müller, Fiscal Policies in a General Equilibrium Model with Persistent Unemployment. VI, 92 pages. 1983.

Vol. 217: Ch. Grootaert, The Relation Between Final Demand and Income Distribution. XIV, 105 pages. 1983.

Vol. 218: P. van Loon, A Dynamic Theory of the Firm: Production, Finance and Investment. VII, 191 pages. 1983.

Vol. 219: E. van Damme, Refinements of the Nash Equilibrium Concept. VI, 151 pages. 1983.

Vol. 220: M. Aoki, Notes on Economic Time Series Analysis: System Theoretic Perspectives. IX, 249 pages. 1983.

Vol. 221: S. Nakamura, An Inter-Industry Translog Model of Prices and Technical Change for the West German Economy. XIV, 290 pages. 1984.

Vol. 222: P. Meier, Energy Systems Analysis for Developing Countries. VI, 344 pages. 1984.

Vol. 223: W. Trockel, Market Demand. VIII, 205 pages. 1984.

Vol. 224: M. Kiy, Ein disaggregiertes Prognosesystem für die Bundesrepublik Deutschland. XVIII, 276 Seiten. 1984.

Vol. 225: T. R. von Ungern-Sternberg, Zur Analyse von Märkten mit unvollständiger Nachfragerinformation. IX, 125 Seiten. 1984

Vol. 226: Selected Topics in Operations Research and Mathematical Economics. Proceedings, 1983. Edited by G. Hammer and D. Pallaschke. IX, 478 pages. 1984.

Vol. 227: Risk and Capital. Proceedings, 1983. Edited by G. Bamberg and K. Spremann. VII, 306 pages. 1984.

Vol. 228: Nonlinear Models of Fluctuating Growth. Proceedings, 1983. Edited by R. M. Goodwin, M. Krüger and A. Vercelli. XVII, 277 pages. 1984.

Vol. 229: Interactive Decision Analysis. Proceedings, 1983. Edited by M. Grauer and A. P. Wierzbicki. VIII, 269 pages. 1984.

Vol. 230: Macro-Economic Planning with Conflicting Goals. Proceedings, 1982. Edited by M. Despontin, P. Nijkamp and J. Spronk. VI, 297 pages. 1984.

Vol. 231: G. F. Newell, The M/M/∞ Service System with Ranked Servers in Heavy Traffic. XI, 126 pages. 1984.

Vol. 232: L. Bauwens, Bayesian Full Information Analysis of Simultaneous Equation Models Using Integration by Monte Carlo. VI, 114 pages. 1984.

Vol. 233: G. Wagenhals, The World Copper Market. XI, 190 pages. 1984.